STRUCTURAL GLASS

PETER RICE
HUGH DUTTON

E & FN SPON
An Imprint of Chapman & Hall
London · Glasgow · Weinheim · New York · Tokyo · Melbourne · Madras

Published by E & FN Spon, an imprint of Chapman & Hall, 2–6 Boundary Row, London SE1 8HN, UK

Chapman & Hall, 2–6 Boundary Row, London SE1 8HN, UK

Blackie Academic & Professional, Wester Cleddens Road, Bishopbriggs, Glasgow G64 2NZ, UK

Chapman & Hall GmbH, Pappelallee 3, 69469 Weinheim, Germany

Chapman & Hall USA, 115 Fifth Avenue, New York NY 10003, USA

Chapman & Hall Japan, ITP-Japan, Kyowa Building, 3F, 2-2-1 Hirakawa-cho, Chiyoda-ku, Tokyo 102, Japan

Chapman & Hall Australia, Thomas Nelson Australia, 102 Dodds Street, South Melbourne, Victoria 3205, Australia

Chapman & Hall India, R. Seshadri, 32 Second Main Road, CIT East, Madras 600 035, India

English language edition 1995

Translated by Martine Erussard

Designed by Richard Medioni

Typeset by Stephen Cary

Printed in France by Mame Imprimeurs, Tours

ISBN 0 419 19940 3

A catalogue record for this book is available from the British Library
Library of Congress Catalog Card Number: 95-68508

ACKNOWLEDGEMENTS

We wish to thank all those who have contributed to the Serres project at La Villette:

The designers
• RFR partners
Peter Rice, Martin Francis, Ian Ritchie
• the team
Henry Bardsley, Guy Deshayes, Hugh Dutton, Isabelle Cacot, Nicolas Hargreaves, Nathalie Mussard, Laurent Aguillon, Gil Besançon, Stefan Reinhart, Gordon Talbot, Étienne Tricot, Jane Wernick.

The participants in the book
Peter Rice, Hugh Dutton, Nataly Bancel, Huguette Boutin, Pat Dallard, Guy Deshayes, Alain Goustard, Thierry Lafont, Jacques Lucan, John Macmin, Birgit Malagon, Claire Mazelet, Guillemette Morel Journel, Philip Moseley, Mariette Muller, Nicolas Prouvé, Ian Ritchie, Emily Stephenson, Daniel Treiber.

The contracting authority
EPPV (Établissement public du parc de la Villette).

The architects
Adrien Fainsilber, architect; ALGOE; RFR; SABOTA; SGTE.

The building control engineers
SOCOTEC

Ministère d'Equipement, Plan Construction, Ruth Marquez, who financed the research and first edition of this book.

The contractors
GIM, general contractors for the Cité des Sciences et de l'Industrie.
CFEM (now Eiffel Constructions Metalliques), main contractors for the Serres.
CFEM's suppliers and sub-contractors:
Boussois, glass; C3F, moulded nodes;
Foundries Manoir, various castings;
Fossier Allard, maintenance equipment;
Masselin, springs; PMB et Balliman, glass manufacturers; Pont à Mousson, tubes manufactured by centrifugal casting; Sarma, glass mechanical suspension systems;
Souchier Fages, ventilation systems; Tecnor, cables;
Voisin, metalwork.

This publication is based on research work carried out by the Plan Lieux de Travail et Constructions Publiques (Places of Work and Public Buildings Scheme), Direction de l'Architecture et de l'Urbanisme (Architecture and Townplanning Directorate), Ministère de l'Equipment.

CONTENTS

DEDICATION

This English edition of *Structural Glass* has been translated from the French second edition, which was completed after Peter Rice's untimely death. We had discussed this project together, particulary the expansion of the book to show how the inital work at La Villette has been appied to other projects by RFR, many of which he had worked on. They examine a theme that was very important to Peter and to the team working with him: that the objective of engineering and technical innovation in building should follow the specific architectural aspirations of each project. I regret that Peter was not able to see this book and wish to dedicate it to him in memory of his tireless and kind guidance, which gave us the confidence necessary to do these projects.

Hugh Dutton
February 1995

FOREWORD

Since the completion of the Cité des Sciences et de l'Industrie, a great number of 'structural glass' applications across the world bear witness to the fact that the aim to create a shop window for new technologies at La Villette has been met.

Peter Rice's inventive capacity during the design of the Serres at La Villette has found many applications that complement his contribution to the evolution of building techniques and to the art of building. Peter Rice's genius managed to persuade the industry 'to think differently about the way to use glass' and to develop the industrial production.

Adrien Fainsilber
Architect
February 1995

FOREWORD TO THE FIRST EDITION

The Serres of the Cité des Sciences et de l'Industrie

The Cité des Sciences et de l'Industrie is surrounded by an urban park situated on a unique site in the capital, crossed by several canals. The architectural solution takes into account the scale of the site and its environment, with the aim of establishing a close relationship between the museum and the park, thus asserting the complementarity between science and nature.

For the park to become a natural extension of the museum, we wanted the south-facing façade to open on to the park and to be as transparent as possible; in contrast, the north-facing façade, on the town side, would be more closed. It was therefore important to control the solar inputs; at the time of the competition we suggested three large Serres (greenhouses) that would create a buffer space, recover the solar energy and integrate the vegetation inside the museum. The bioclimatic façade, which we wanted as transparent as possible, would be the link between the museum and the park, and the glazed lifts within this 'double skin' would enable visitors to view the park.

Peter Rice and his team were able to translate the desired architectural composition in masterly fashion by building the façades with a minimum of 'material', just as the builders of the Sainte-Chapelle had done. The preciousness of the assembly systems blends with the sobriety of the composition to give an impression of monumental lightness and purity. The prestressed cable trusses are laid out horizontally, so that the transparency is almost total at eye level. The external surface is smooth and continuous. During the design stages under the leadership of Peter Rice and his associates, the structure of the Serres was made lighter and lighter, until it became almost immaterial. I wish to emphasize this fact because it is exceptional: indeed, the stresses applied generate, more often than not, a redundancy in the structural steps necessary to meet safety requirements.

The Serres at La Villette are an inherent part of its permanent structures which illustrate new technologies; they contribute to the Cité des Science's strategy of enabling the French people to discover the great industrial adventure of today and tomorrow and helping them to understand better the current developments at the threshold of a new industrial era, influenced by the computer and audiovisual sciences, and to further creativity and innovation.

Adrien Fainsilber
February 1990

THE SERRES AND THE CONCEPT OF TRANSPARENCY

Little story of a big project

The creation of a National Museum of Science, Technology and Industry at La Villette was first suggested in 1978 by the French President, Valéry Giscard d'Estaing. To this end it was decided to convert the site of the disused abattoirs (built in the 1960s but never completed), auction and market halls into a museum surrounded by a park. The idea was to create a museum lying conceptually somewhere between the Science Museum in London and the Air and Space Museum at the Smithsonian Institute in Washington: Paris had no equivalent at that time.

A limited competition was held in 1980. The winner was Adrien Fainsilber with a scheme to renovate the abattoirs and to surround the main building with moats and a park. The geometric design of the park placed the museum in a formal setting that included water, an important element in the design. In this scheme, three of the four bays facing the park were to have a large glass façade, one to the west and two to the east. The fourth bay was to become the entrance area. A strong theme in the competition design for the museum was that it must express such elements as light, water and solar energy. The structural design of the three main façades, or Serres, as they are called in French, was such as to enable the inclusion of these elements in the buildings and to enhance the relationship between the museum and the park. An internal Serre transferring light downwards into the lower part of the building was also included, but subsequently dropped during the evolution of the design.

The roof of the building had 12 domes, 18 m in diameter, crowning each bay of the building. These domes were modelled on the glass dome of the Johnson Wax headquarters (Frank Lloyd Wright, Wisconsin, USA, 1936–1939) and permitted the whole of the top floor of the building to be lit from above. During the design phase, the 12 domes were replaced by two above the central entrance area. Many changes were

Model of the competition entry (1980) of the National Museum of Science, Technology and Industry by Adrien Fainsilber.

made following the competition and also at the request of Valéry Giscard d'Estaing himself. For example, he expressed a wish that the Géode be incorporated within the building and therefore the lighting from the top of the entrance area had to be abandoned, as the Géode was described as an 'ungraceful protrusion contrary to the French taste'. However, these changes were also a direct response to the requirements of the museum curators; indeed, as the project progressed, they were able to define and to detail the utilization of the spaces, as well as the nature of the relationship between the building and the surrounding park. Despite the demand that the Géode disappear in the building, it was later moved back to its initial position.

The main change that occurred after 1981 was the abandonment of Adrien Fainsilber's design for the park, which became the subject of a separate competition. The competition brief and the winning project, designed by Bernard Tschumi (in 1983), kept the moats and the design of the immediate vicinity of the building of the first project. The building itself was to remain unchanged; the Serres were also to remain unchanged both in design and architectural function. The spherical Géode, clad in stainless steel opposite the museum entrance, was to become one of the principal images of the building together with the Serres and the two domes of the entrance area.

The project architect, Adrien Fainsilber, defined the Serres as a transition space between the building and the park and wanted them to be as light and transparent as possible, on the model of the state of the art glass technology used for the headquarters of the insurance company Willis Faber & Dumas (Norman Foster, Ipswich, UK, 1975). The technical character of the Serres, which in this instance bears no association with the architectural 'style' called *high-tech*, was to become stronger and stronger as the project evolved. The use of glass and its innovative implementation expressed, in terms of building construction, the nature of the museum and of its exhibits. This same technical character was later to become a predominant inspiration for the design: to create a building whose elements reflected the best of current French building technology.

Shortly after winning the competition, Adrien Fainsilber invited Peter Rice to work with him. Rice, who had just ended his partnership with Renzo Piano, accepted and started work on the Serres' structure, on the domes and on the internal dome as first planned. Indeed, these three projects were three independent architectural entities, each demanding its own detailed design, and they could be worked on in direct consultation with the architect. They did not need the large team that would have been necessary to work on other elements of the entire building.

Peter Rice, who was aware of Martin Francis's experience of working with glass and that he was resident in France, asked him to form a team. Together they invited the architect Ian Ritchie to join them.

Peter Rice was a structural engineer by training and worked in this field for many years. He was involved with the Sydney Opera House (Jørn Utzon, 1956–73) as consultant and site engineer and with the Centre Pompidou (Renzo Piano and Richard Rogers, Paris, 1977) as chief engineer and designer of the steel structure.

Martin Francis designed the glass façade for the Willis Faber & Dumas office block. For a few years now he has worked as a naval architect on the Côte d'Azur, specializing in large sailing and motor yachts.

Ian Ritchie is an architect. He met Martin Francis while working on the Willis Faber & Dumas project and has also worked with Peter Rice at Ove Arup and Partners in London.

The different training backgrounds and skills of the individual members of the team gave shape to the design of the Serres. The final design is the result of a close cooperation between Adrien Fainsilber and the RFR team, which during the project phase was reinforced by Henry Bardsley (an engineer who had worked with Peter Rice at Ove Arup and Piano + Rice) and by Hugh Dutton (a young architect from the Architectural Association School in London).

Each architectural project is, before anything else, the display of a collective effort. In the case of the Serres, in particular, the project as a whole is far too complex to be handled by one person, whether the architect, the engineer or the contractor. Mutual trust and communication are essential and only the pooling of the individual expertise and skills can allow the achievement of a project of quality. There is no doubt that the story of the Serres could be retold differently by each member of the team; nevertheless, we shall try to relate everyone's contribution.

Within the design and construction of a building, the architect has a key role to play. It is the architect who is the project team leader, who defines the concept and the priorities, and sets and reinforces the aesthetic objectives and the propositions that are acceptable. All technical solutions must be approved by him or her. Constant communication between the architect and the other team members is therefore essential to ensure that all the facets meld together and contribute to the creation of the whole, and that the functional requirements are clearly defined within the building.

This short history of the project is intended to demonstrate, on the one hand, how this communication worked, and, on the other, the scale of such a project and the need for trust and real cooperation between industry, the designer and the client. In the case of the Serres, nothing could have been achieved without the active participation of the glass and fabrication industries – and this from the very beginning of the project.

The footbridge spanning the moats provides access to the museum's entrance area between two Serres.

The fact that this account relates to the Serres is not because of their dimension or 'spectacular' quality, but because they represent a specific aim whose objectives are clearly defined and whose performance can easily be demonstrated.

The complexity of the work involved in the Serres is also to be found in the whole structure of the museum.

Top:
Methods of fixing glass in the Middle Ages.
Bottom:
West windows (twelfth century) in Chartres cathedral.

The scale and diversity of the design difficulties raised reflect the make-up of our society and the depth of knowledge needed to create something new. Every single aspect must be controlled and motivated and this is where the architect and other projects leaders play their part.

The aim of this book is to present the thinking that lay behind the project of the Serres. The following paragraphs will place the project in the historical context of the architectural concept of transparency. Then the Introduction will present the method and philosophy which guided the project and whose key elements are: predictability, process and hierarchy.

Transparency through time

Certain periods in the history of glass are particularly relevant to the concept of architectural transparency. These will be discussed from three different viewpoints: one-way transparency, two-way transparency and the expression of the transparent surface itself. Why this historical glance at transparency? Because it recalls and illustrates one of the fundamental objectives of the Serre. In describing the various types of transparency which the ancient glass manufacturing techniques permitted, it is hoped that the reader will gain a more precise idea of what the design team actually understood by 'transparency'.

Even before the advent of glass, the very first architectural use of windows found its origin in the need to allow daylight to penetrate to the interior of buildings. Here lies the definition of one-way transparency; where the glass serves as nothing more than a means of preventing the rain from getting in and the heat from getting out.

The concept of two-way transparency appears as soon as the view to the outside acquires an importance: for instance, the view of a garden laid out for this purpose becomes a deliberate element of the architectural composition. Finally, when the very nature of the transparent surface is an integral part of the actual composition, particular attention is drawn to the glass itself as a material and to the way in which it is supported.

One-way transparency

In earliest times, openings were created in the walls of buildings to allow daylight to enter. Later, the use of glass for the windows allowed the penetration of daylight while affording protection against the weather. The early glass-making techniques of the Middle Ages

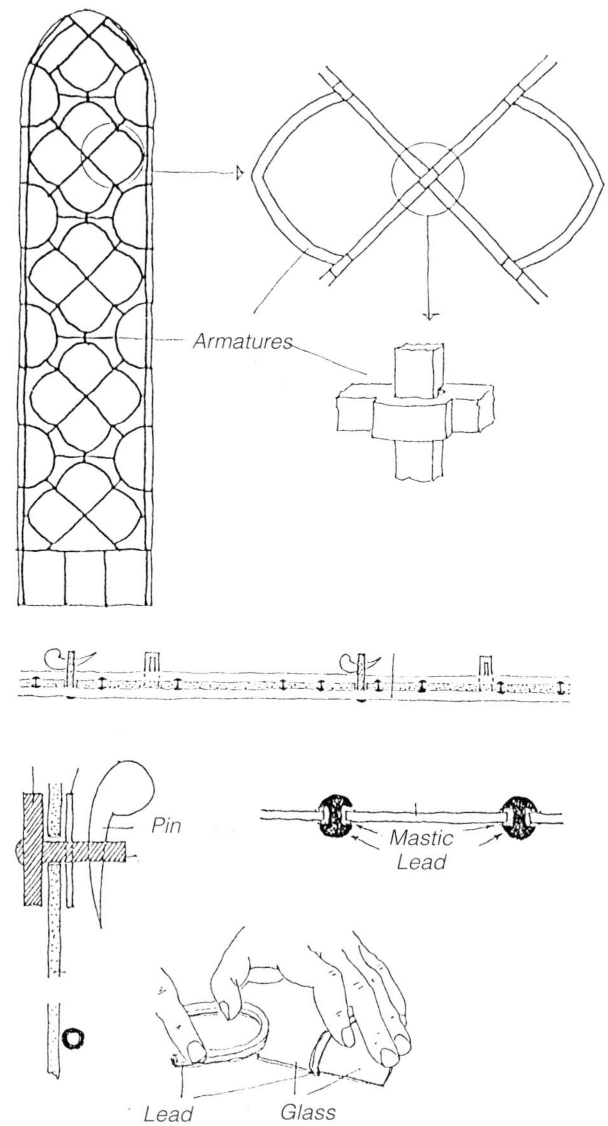

Top:
Gothic window in a house at Béziers.
Bottom:
Making flat glass at the end of the seventeenth century.

placed limitations on the uses of glass. With such techniques one could only obtain cut pieces of a very small size, demanding in their turn a very dense supporting structure. The framework was thus an important part of the window design and the aesthetic potential of the lattice work of the frame, seen in silhouette from inside, was exploited. Glass panes were held together with 'H' shaped lead sections and sealed with an oil based mastic. Furthermore, glass made in this way had poor light transmission qualities; the surfaces were not smooth and the material had many small imperfections and discolorations. The two technical constraints – size and optical quality of the glass pieces – limited the potential of a one-way transparency window.

Two-way transparency

With progressive improvements in glass-making techniques, a new dimension, the 'view out', could be considered. The clarity of the image perceived from the other side of glazing depends firstly on the smoothness of the glass surface, secondly on the size of the glass pieces and thirdly on the purity of the material; that is to say, the absence of discoloration, air bubbles or impurities.

In the seventeenth century, the manufacture of flat sheets of glass was able to meet the increasing demand for transparency. The process commonly used at the time consisted of pouring the molten glass onto a flat preheated metal bed and then flattening it with a roller until an even thickness was achieved. However, the rollers and the bed left marks on the surface and, as a result the sheets of glass were at best translucent and had to be painstakingly polished to achieve a perfectly smooth and transparent surface. The transparency of the panes of glass enabled the architects to include the view out in their designs. The external environment was therefore able to play an important part in the composition of the internal spaces.

The advent of opening windows helped the communication between the inside and the outside; and transparency acquired a new dimension – the two-way dimension. The palace and gardens at Versailles (*c.* 1670) are such an example. At the time Versailles was built, glass windows were widely used and were an essential element of architecture. The relatively large glass windows were held in wooden frames mounted on hinges. The frames had ornamental mouldings on the inside and a small rebate on the outside. The glass was secured in the frame by a bead of mastic, which

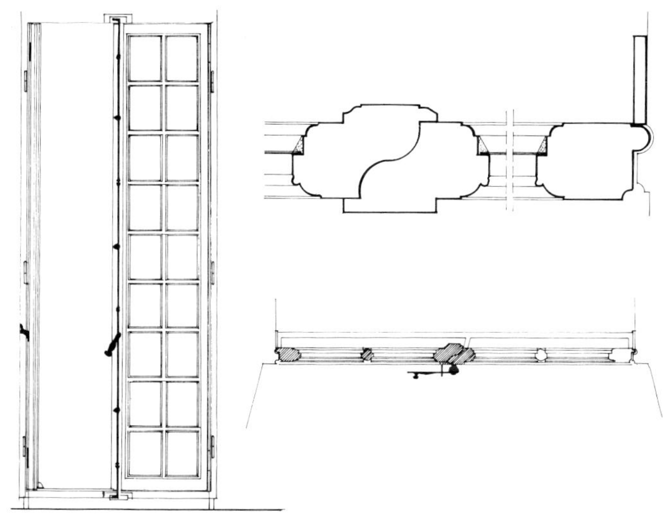

Chateau de Versailles.
Top:
The marble court.
Bottom:
The joinery of the bays.

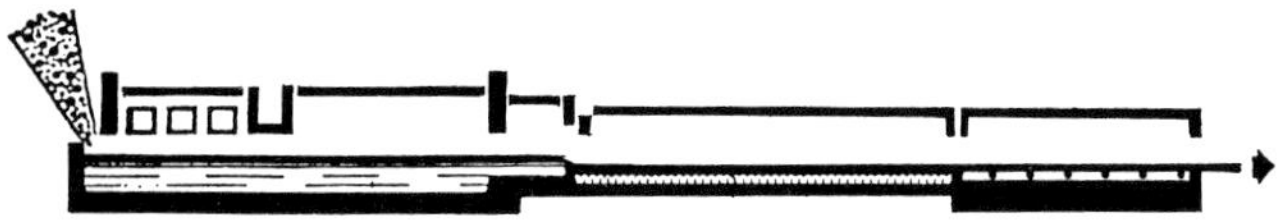

Principle of the Pilkington float system.

How transparency plays between the spaces making up the promenade architecturale created by Le Corbusier in the Villa Savoye.

Through the use of a large pane of flat glass, the landscape penetrates the interior of the House at Seeburg (Hans Hofmann, Lucerne, Switzerland, 1950).

The large glass surfaces, held by woodwork of minimal section, create an impression of spatial ambiguity between the interior and exterior of the Farnsworth house.

also served as a waterproof seal – a detail which is nowadays commonplace.

As glass technology developed, the sheets of glass became larger and architecture exploited this new dimensional liberty.

At the time of the Modern Movement, the use of large plate glass sheets considerably reduced the supporting framework and in some cases eliminated it altogether. Where necessary, the glazing bars could be of metal and therefore very thin.

The industrial manufacture of glass dates back to the beginning of the twentieth century. The Fourcault process (patented in 1904) involved drawing a continuous band of glass vertically from a molten bed using asbestos-covered rollers. At almost the same time, the Libbey-Owens process (patented in 1905) improved on this method by drawing the glass horizontally. These methods have been refined throughout the century and most significantly by the Pittsburgh Glass Company (PPG). This company invented a process whereby the glass was drawn by air-cooled rollers and the glass sheet held only by the edges, thereby eliminating any imperfections in the surface of the glass that may have been introduced by the rollers.

The invention of the Pilkington float system in 1952 revolutionized the production of glass. Molten glass is poured onto a bed of liquid tin and spreads out and floats like oil on water, forming an even film. It is then drawn as a continuous band by rollers. Since there is no imperfection in the metal surface, the glass surface does not need to be polished.

The architects of the Modern Movement exploited these technical advances to create two-way transparency designs which became progressively more impressive.

The transparent corner of the Bauhaus building (Walter Gropius, Dessau, GDR, 1925–26) is a celebrated illustration expressing the idea of the view through the building.

The quest for light and airiness of spaces and for a closer contact with nature was another important theme of the Modern Movement; the Farnsworth House (Mies van der Rohe, Plano, Illinois, 1946–1950) is a typical example of this idea.

The concept of large uninterrupted surfaces of glass is achieved in a most spectacular fashion at the 'Maison de la Radio' (Henri Bernard, Paris, 1953–1963). However, a very large sheet of glass of the size used in that building is subject to bending (due to wind loads) and to buckling (due to its own weight). The Hahn system, adopted here, resolves these two problems in a straightforward manner: verti-

Above:
The transparency of the corner of the Bauhaus building.
Below:
The Hahn glass suspension system.

cal glass fins are fitted perpendicular to the surface to give the required bending stiffness. The glass is then simply hung as a curtain and is therefore always in tension, which removes the necessity for resistance to self-weight-induced buckling. The glass is held by a series of clamps which are fixed to rocker arms in order to ensure an equal load distribution across all of them.

These examples illustrate the increasing fulfilment of the desire to create a weather impermeable surface which simultaneously suggests an invisible plane, letting the light in and allowing a view out.

Expression of the transparent surface

A third aspect of transparency is linked to the presence (or the 'non-presence') of the surface through which one looks. Transparency implies looking through something; the viewer must therefore be aware of the glazed plane through which he or she is looking. In some cases, architects make a particular effort to

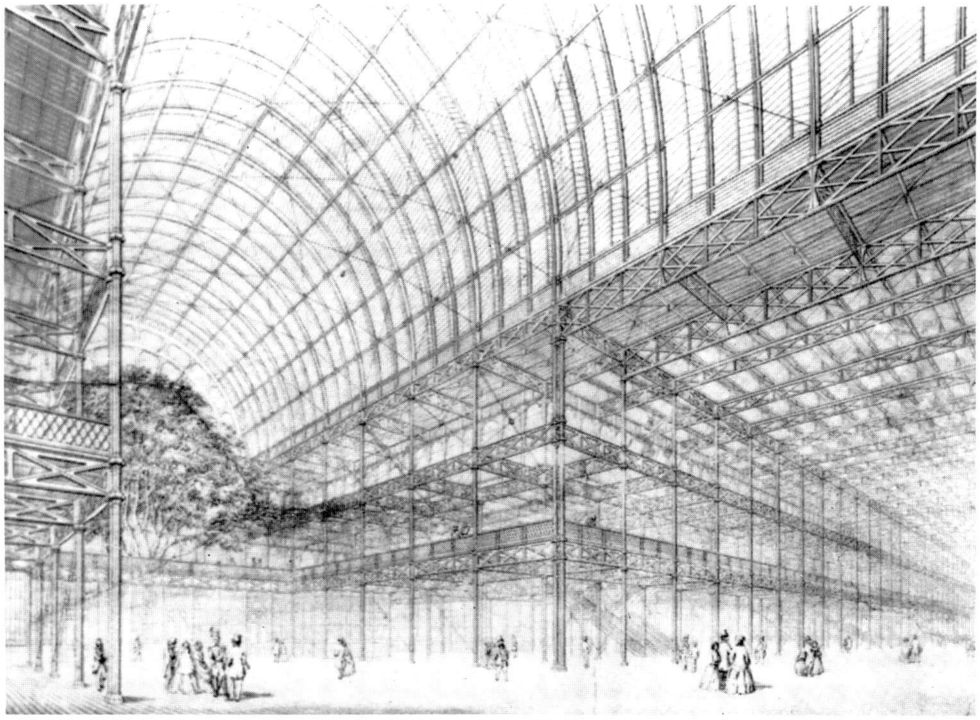

Crystal Palace.
Left:
View of the interior before mounting the exhibits (towards the end of 1850).
Right:
Changing the glazing panels in the transept in 1899.

exploit the sculptural or compositional potential of the glass surface itself and of its framework. This notion has long been in existence, as is demonstrated by the windows of Gothic cathedrals. The following examples are particularly inspiring exercises in transparency.

The Crystal Palace (Joseph Paxton, London, 1851): the vast glass nave illustrates the idea of spatial transparency. The first industrial achievements of mass production and the technical feats which they represented were on show here. Infilling glass panels were fixed on a lightweight metal frame and built with remarkable speed. The glass itself did not have a structural role.

The Glass Pavilion (Bruno Taut, Cologne, Germany, 1914), which housed the glass industry exhibitions, illustrates the same idea but on a much smaller scale. It is a didactic representation of all the glass materials then available: the walls, the floor, the dome and even the staircase are made of glass. The glass is used structurally as a building material and is therefore load-bearing. The interest of such an achievement lies in the intention to place the visitor in an all-glass space in order that he or she may perceive the subtle modulations of the building's outer skin by natural light. Here transparency becomes three-dimensional.

The Villa Savoye (Le Corbusier, Poissy, Yvelines, 1929–31) explores transparency and its interaction with spatial composition. The large curved bay on the ground floor follows the rhythm of the vertical mullions spaced every 22 cm. The result is a very interesting non-planar transparent surface. The repetitive mullions define the curved surface in space, this curve being both present and absent (because of the transparency of the glass). The rest of the villa expresses an ambiguity between inside and outside; the long windows are continuous around the entire first floor, even on the terrace where there is no glass in them. The interpenetration of the spaces is particularly noticeable between the living room and the terrace.

The final example is the Palm House (H. and D. Bailey, Bicton, UK, 1843). It is a greenhouse with a considerably large central volume and two wings. The transparent surface, the glass and glazing bars, form a very delicate skin, the primary supporting structure being reduced to the absolute minimum. Very small fragments of glass are fixed onto curving glazing bars such that they make a smooth rounded shape. The glass itself is the only means of lateral bracing for the glazing bars and therefore plays a structural role. Together the glass and the bars form a transparent shell similar to that of an animal, and this shell form provides its structural stability. Minimizing the visual impact of the supporting elements maximizes the one-way and two-way transparency. At the same time the transparent surface finds its own expression through the particularly economical use of the glass fragments

The greenhouses of the Palm House at Bicton.

Bruno Taut making exclusive and didactic use of glass in the pavilion built for the Cologne exhibition.

and of the small metal bars of this smooth and gently curving skin.

Transparency and reflectivity

The transparency of glass depends on the relative amount of light on either side of its surface. If there is more light on the viewer's side than on the reverse side, then it becomes reflective. If, on the other hand, there is more light on the reverse side, then it will be transparent.

This can be clearly seen in Jean Nouvel's building for the Fondation Cartier, in Boulevard Raspail, Paris (completed in 1994), where the sky is the dominant light source. Often when large glass planes are used in buildings, you can see in only with difficulty, or when the lights are on inside. This is because there is always more light outside than in. Furthermore, there is always more light in the sky, particularly on an overcast day, than at ground level, where it is absorbed by the surroundings. For this reason, when you look at a glass plane at an angle such that the sky is reflected,

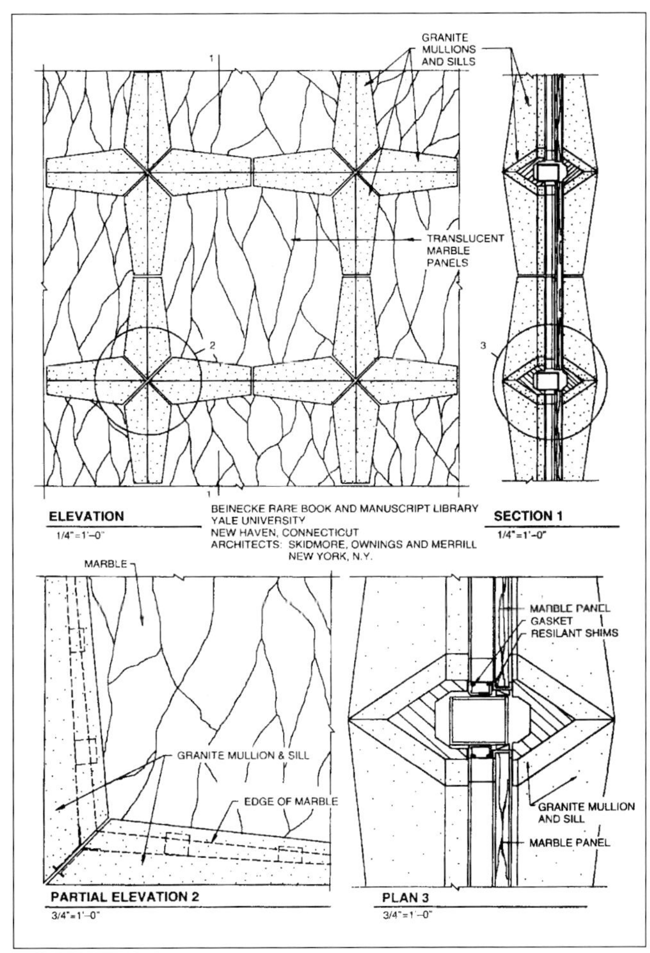

Right and following page: The translucent marble panels of the Beinecke Rare Book and Manuscript Library at Yale University are held in place by mullions of preformed concrete that give a vigorous modernity to the interior and exterior façades.

for example on the Grande Pyramide at the Louvre, it is not totally transparent.

The luminous plane in architecture

In addition to the transparent and reflective plane of clear glass, it is also interesting to consider the role in architecture of the luminous surface. This is another architectural device that involves the use of glass or other translucent materials. Marble and alabaster have long been used in architecture to create light surfaces. They occur, for example, in the old St Peter's Church in the Vatican (*c.* 300–320 AD) and more recently S.O.M. (Gordon Bunshaft) used marble for the Beinecke Rare Book Library at Yale. In the 1980s Arata Isosaki used onyx in his Museum of Modern Art in Los Angeles. Recent advances in stone-cutting technology have now made it possible to laminate a fine film of marble, as thin as 1 mm, on to glass, thus making a particularly luminous and rich surface.

In the pure glass field, translucent glass has existed for a long time. It can be sand blasted or included as a translucent interlayer in laminate products. A recent fashion in architecture is 'fritting', which produces a pattern of paint enamelled on to the glass surface. Norman Foster's work at Stockley Park (1989) is one of the earliest examples of this technique.

The future of the plane of glass in architecture

The techniques of fixing glass and exploiting its structural capacity that are discussed in this book allow structural elements to be eliminated from the glass plane. This has a particular potential in architectural expression, because until recently the transparent or luminous surface has been dominated by the mullions or glazing bars that have been necessary to hold the glass in place.

The flimsiest steel or aluminium profiles, and even glass stiffening fins, detract from the pure planar nature of the glass surface, giving it a relief standing out from the reflective or transparent planes. Removing them creates a new architectural device – a pure planar surface, perhaps 20 or 30 m high of transparent, reflective or luminous glass. Some of the most recent examples of the applications of this technology, such as the glazed atrium for 50 Avenue Montaigne in Paris (described in the Postscript), are beginning to explore this architectural potential.

Preceding page:
One of the Serres at the City of Science and Industry, Paris, 1986.
Left:
Banque Populaire de l'Ouest et de l'Armorique, Montgermont, 1990.
Right:
The 'Grande Nef' of the Collines du Nord, La Défense, Paris, 1990.

Left:
Private club, Tokyo, 1992.

Above:
Large and serial greenhouses at the Parc Citroën, Paris, 1992.

Above:
Inverted pyramid at the Grand Louvre, Paris, 1993.

Right:
TGV railway interchange, Roissy, 1994.

DESIGN PHILOSOPHY AND METHODOLOGY

Transparency

From a spatial point of view, the ambition of the Serres project was to maximize one-way and two-way transparency, thereby rendering the inner space as full of light as possible, giving primary importance to the views from outside, from inside and through the volume of the building itself.

The basic architectural objective defined in the competition scheme was that the Serres be transition zones between the museum and the park. The spatial transparency of the Serres is exploited in an architectural composition of conscious spatial ambiguity, creating a place where one is neither inside nor outside, but is simultaneously in the museum and in the park.

The glass and glazing system are designed to make the public subtly aware of the transparent plane it is supporting. It attempts to remind the viewer of the presence of the plane of glass, while at the same time maximizing its ephemeral and transparent qualities. From inside, the eye recognizes a series of points and lines (the supports and cables) which define a plane in space; from outside, the glazed plane is materialized in the form of a perfectly smooth and uninterrupted reflective skin.

Thus the view from inside is dominated by the glass-supporting structure: the glass itself is as if dematerialized, because it is perceived only through its reflection. Since the internal light is more subdued than the outside light, the glass is no longer visible from inside because its reflectivity is overpowered by the light coming in. Furthermore, all the support system details are designed to enhance the inherent flatness of the glass surface.

In contrast, the view from outside is dominated by the reflectivity of the light on the glass, since the light reaching the glass from outside is stronger than the light coming from inside the Serre and the museum. Thus the presence of the glass is perceived through its reflective capacity. All support details are designed not to compromise the smoothness of the glass, so that it remains a continuously readable surface: the bolts are all countersunk and no element protrudes from the external glass skin.

The total transparency of the glass envelope of the Serre means that light can penetrate at many different angles. This means that when seen from outside, the amount of light behind the glass is always quite considerable. Furthermore, the light falling on the glass is never absolutely reflected, as it would be in a mirror, and thus the space tends to adopt an immaterial quality. The amount of light coming from either side of the glass varies quite extensively and a constant opposition effect between immateriality and reflectivity of the glass surface is thus created.

As we said before, three concepts – process, predictability and hierarchy – are the key conceptual bases for the suspended façades project of the Serres at La Villette.

1. Process

It is important to understand the design process used in the Serres project to appreciate how it works. This requires a clear understanding of the structure's behaviour and its architectural objectives, plus a knowledge of the possibilities made available by industry. Such constraints demand a rigorous method of working and follow-up.

The Serre is not simply a good idea conceived and then completed in the abstract. It is the result of long research into the behaviour of glass, as well as the fixings, the cables and the structures that support it. The project evolved as understanding grew wider and deeper. The results of each design stage were compared with the aesthetic and architectural objectives and the necessary decisions taken accordingly. This type of design process is somewhat analogous to that for industrial products such as aircraft, motor cars, machinery and even skis.

The fact that the design process is a combination of research, information, current practice, experience and

A glazed plane is transparent or opaque and reflecting depending on the angle at which it is viewed: it is always vacillating between the immaterial and the material.

application of aesthetic objectives is typical of our time and of the nature of modern industry. Far too often, the aesthetic objective is reduced to a superficial veneer on a product primarily designed to fulfil its function correctly.

To achieve a better integration between the aesthetic and functional objectives, some or all of the behaviour must be understood and handled by the designer. Architects and engineers working in the building industry are in a unique position to achieve this fusion: by custom and by law they control the projects – buildings or construction works – for which they are responsible. They have the skills to understand and to process the information necessary to control the design and the implementation of a given project.

But when it comes to its execution, the initiative belongs to the manufacturers and contractors. This does not imply that industrial products are bad – they may be functional and even aesthetic. However, the objectives of industry are different from those of the designer and are often too conventional and limiting. Great designers of the past, like Jean Prouvé and Pier Luigi Nervi, joined with industry and created within that very context. The design and building of the Serres is a small example of an alternative approach.

There is no doubt that industrial concerns are willing to commit themselves and are able to be cooperative. They are often keen to encourage adventure and experiment. But right from the start they must be convinced of the feasibility of the project. Their scepticism can be explained by the fact that they, too, will be responsible for everything that will be built. In the case of La Villette, CFEM (now called Eiffel Constructions Métalliques), Boussois and the other contractors involved, as well as Socotec, the 'Bureau de contrôle', played an important part in the development of the project from the very beginning of their involvement. They had accepted the challenge and, once involved, they cooperated positively towards its completion. Examples of positive cooperation between designers and industry are numerous and we shall use a few of them to illustrate this synergy in more detail.

- Detail of the holes in the glass and the improvement of the drilling process

CFEM and Boussois worked together on the testing process to demonstrate the feasibility of the system. In the course of testing, considerable improvements were made to this particularly critical part of the glazing system. Visits to the Boussois factory in Bobigny resulted in a close cooperation between the project team and the manufacturer which led to the design of a special drilling jig (Chapter 1, p. 40).

- Detail of the glass support system

The firm Unibal, manufacturer of standard spherical bearings, were contacted at the beginning of the project and it was their first prototypes that convinced us of the feasibility of this component. Then Sarma, an aviation equipment manufacturer, undertook the detailed development work on the components and finally produced them (Chapter 1, p. 37 onwards).

- Castings in the glass support system

The idea for these parts came following a visit to Le Bourget Air Show and after having contacted ATS France SA, manufacturer of components for the aviation industry. This firm contributed its know-how and expertise to the design development of these pieces. Here again visits to the factory and the understanding of the manufacturing process contributed to the improvement of the parts (Chapter 2, p. 55 onwards).

- The casting of the main frame node

Fonderies de l'Acier de l'Est helped the design team a great deal in understanding the technical requirements of the casting process in relation to the required shape of the part to be made. Eventually the contract was awarded to the casting works C3F, which conducted visits around their plant so that the team could better understand the intricacies of their manufacturing processes: comprehension of the methods made evident which shapes could actually be produced. All this information has greatly influenced the design. Furthermore, the company made use of computer programs to generate complex finite element analyses of the cast shapes. This company undertook the detailed design of the part and made all the necessary final adjustments (see Chapter 4, p. 86 onwards for a detailed description of the castings).

Generally speaking, this interactive approach resulted in some strange 'partnerships' between companies that had never before worked together. In this way glass manufacturers came into contact with the world of aviation and with large mechanical and contracting industries.

These examples of positive cooperation between designers and industry do not lessen the need for rigorous control on the part of the architect during all the design stages. The only negative aspect of this type of cooperation lies in the contractual system that governs the building industry. A standard contract requires a clear and detailed description of the work to be carried out. The manufacturer does not receive any payment

The Serres seek to make the public aware of the existence of the transparent plan: the glass-hanging system and the array of point supports define the invisible plane in space.

The transparency of a Serre used to advantage in the outlook on a spatial ambiguity: it contributes to a definition of a place where one is neither inside nor outside, but in the museum and the park at the same time.

for his involvement in the design before the contract is signed, and once it is, all amendments or improvements open the way to additional payments and other extra costs. The contract must be written in such a way as to give a certain latitude for development without causing financial surprises.

2. Predictability

To understand fully the concept on which the Serres are conceived, it is necessary to understand the physical and structural characteristics of glass. A very simple logic links the solutions finally adopted to the physical properties of the glass. Nothing in the solution is arbitrary. Each joint, each bearing, reflects the way in which glass behaves. And, therefore, apart from emphasizing the idea of transparency and creating an immaterial space, the guiding principle of the Serres is 'predictability'.

Each part, each detail of the structure has been chosen so that its behaviour under load can be predicted exactly. The design tries to indicate, as far as possible, how the loads are transferred into the various parts of the structure, to isolate the functions particular to each element and to allow the behaviour of each assembly to be understood. This approach facilitates the 'predictability' – so important with a material as fragile as glass – of the performance that can be expected from the system. Should a glazed unit break, it is then easy to calculate analytically the loads thus created. The principle of predictability can be applied to any glazed surface design; it entails removing all uncertainty about the behaviour of the design by eliminating those elements whose structural behaviour might be ambiguous or unclear. This principle is particularly relevant to the structure of the Serres: glass is a material that carries load in a simple linear fashion up to the threshold at which fracture occurs, after which there is failure of an almost explosive nature.

The more common building materials such as steel, timber and concrete show some resistance to crack growth. Small fissures or flaws can occur within the material without causing failure because their structure is resistant to crack propagation. But glass does not have these properties. In design terms, this means that one has to be sure of the loads and stresses to which the glass is being subjected.

Because of their ductile behaviour, common building materials are very tolerant to high local stresses and discontinuities in their support; this means that steel assemblies can be rigidly bolted since some plastic flow in the material will relieve any high stresses that may occur.

None of this is true of glass, where any stress or distortion that exceeds the linear strength will cause brittle failure. In traditional glass construction, this problem has been resolved by adding a flexible and absorbing substance, such as a rubber seal or pad, between the glass and its support.

The approach adopted in the Serres' design was different. The support system was conceived so that there was always a clear analysable load path. The stresses in the glass have been very precisely predicted under all load conditions. The design of the spherical bearing assemblies at the glass support points and of the compressed spring support system are two examples of the application of these principles. The requirements for predictable behaviour under load were used in a positive way to express the inherent physical characteristics of the glass and to enhance the transparency and lightness of the whole.

3. Hierarchy

A third concept employed in the design of the Serres is that of 'hierarchy'.

In a structure the notion of hierarchy indicates that all the elements that constitute the structure lie in a naturally correct relationship with one another. In the case of the Serres at La Villette, this means that each part, each section, can be immediately identified as being an integral part of the logical reasoning behind the whole building.

Natural structures possess this same property. Each part of an oak grows according to the same hierarchical logic as the rest of the tree. It is only the scale of the phenomenon that varies. The same rules apply both on a 'macroscale' and on a 'microscale'. It is not difficult to understand why this should be so in our natural environment, and it is important, but not so easy, to remember these rules and relationships in our built environment, particularly when referring to a simple and monofunctional structure like a Serre. But these rules must be deduced in order to understand how the hierarchy and the cohesion of the structure work.

In the case of the Serres, one rule is the need for predictability. This means that all details of the various elements must be designed and drawn in order to ensure that the elements under load will always be the same in relation to each decision taken with respect to each detail.

Another aspect of this quest for hierarchy concerns differences of scale. Some elements of the whole carry heavy loads over a long span. Others carry small well-defined segments. They must not be confused. Both exist and must be understood according to their functions, while being perceived as part of the same logic of expression.

The recent mathematical studies of fractals show that this phenomenon may contain a fundamental key to the natural environment and to natural structures. The decreasing sequence of the elements of a structure forms an entirety carefully put together and in which the smallest and the largest elements are directly linked. The rigorous exclusion of any superfluous or inadequate element or material is an important stage in the search for a hierarchy that has been correctly achieved.

The key consideration of the Serres is that of the structure. It connects all parts together and therefore controls the form. The second consideration is that of tension: tension in the wind-bracing system, tension in the structure, tension in the glass, all implying a way of bearing the loads that clearly expresses the cohesion between the parts and the whole. The design flows entirely from these two considerations. These are the rules that are the basis of the structural hierarchy of the Serres.

The structural composition follows a rigorous hierarchical sequence. The projected elements are classified below in decreasing order, from the whole to the detail:

- the primary tubular structure frame
- the cable truss
- the glass suspension system
- the glass and its support points.

No element is affected by the behaviour of the preceding one in this hierarchy. By contrast, each element bears the load of those subordinate to it according to this hierarchy, and must therefore be able to take the stresses created by them under varying external conditions.

Thus all the elements of the hierarchy serve the plane of glass and its supporting elements; they are defined by the glazing unit that they support. It therefore seems that the description must start with this, at the bottom of the hierarchy.

1. The plane of glass, itself made perceptible by its reflections, is held by an array of identical support points. These details are characterized by an oversized washer which renders them visible and gives a presence to the plane of glass.
2. The glass suspension system is detailed so as to give the effect of a curtain of suspended glass. The suspension spring and the vertical hanging connections in the H-shaped support assembly are both clearly expressed.
3. The cable truss system reduces to a minimum the wind-bracing for the glass; therefore the viewer sees the plane of glass, not the busy grid of glass mullions used in conventional systems.
4. The tubular structure frame is an 8 × 8 m grid placed just inside the plane of glass. Made up of tubes 300 mm in diameter, this grid is braced by a thinner bracing truss. The technique of prestress, described in detail later on, allows this bracing truss to consist mostly of tension member rods between 30 and 55 mm in diameter. The relative thinness of the bracing trusses allows the tubular grid to retain a distinct expression.

The following four chapters describe each of these elements in the order used above.

The design of the glazed plane rigorously respects the concept of hierarchy: from the glass and its fixings (Chapter 1) to the suspension system (Chapter 2), to the cable truss system (Chapter 3) and as far as the structure (Chapter 4), each element is immediately identifiable as making up an integral part of the complete work.

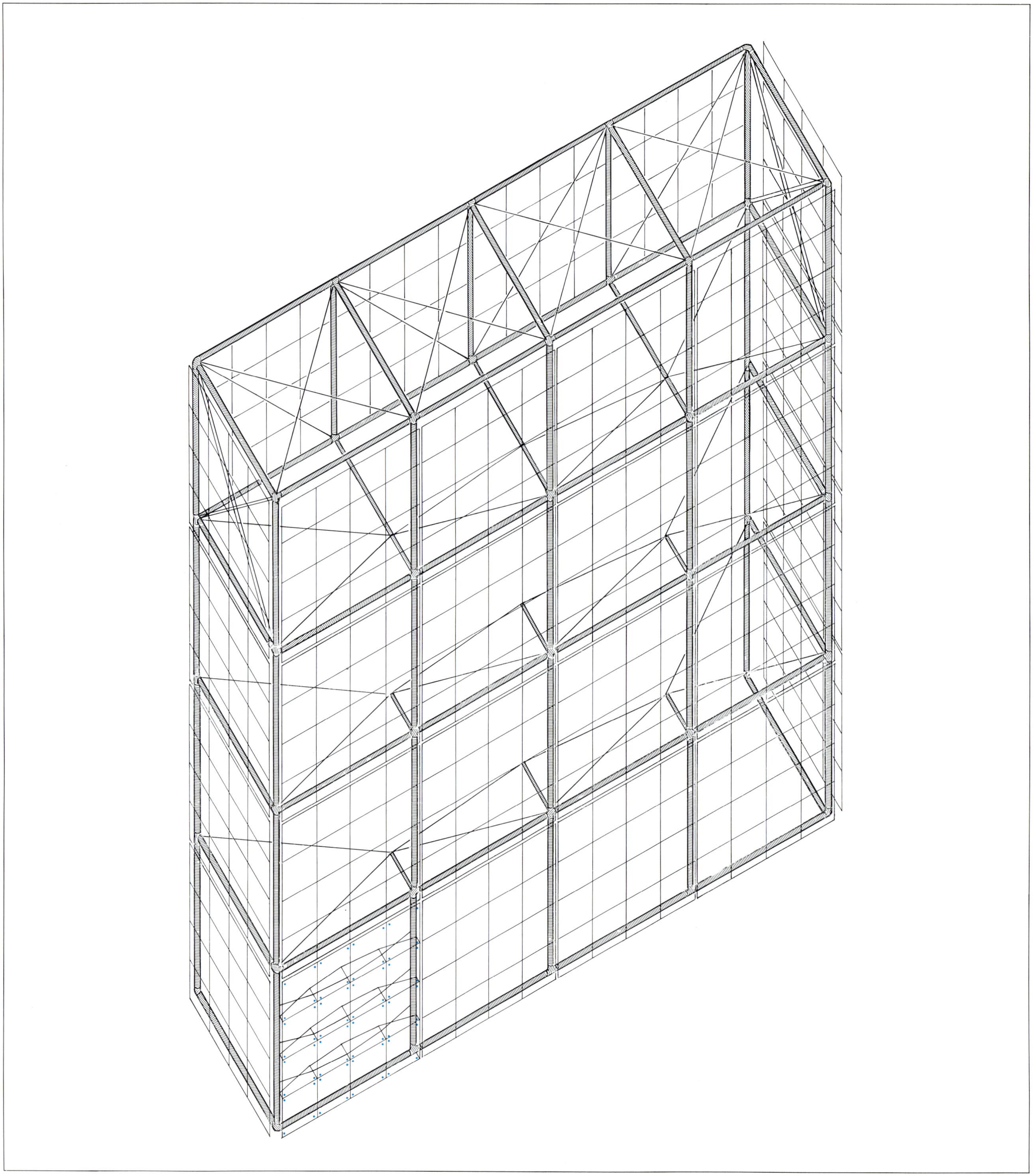

CHAPTER 1

THE GLASS AND ITS SUPPORT POINTS

The properties of glass

Materials are rarely perfect, they are likely to contain small crack-like defects. When a crack of a given size is loaded beyond a certain critical stress, the crack will propagate and failure by fracture is likely to result. The material property that relates critical stress to the size of the crack is the fracture toughness. Steels have a relatively high fracture toughness. For typical defect sizes, the critical stress for a steel component is usually well in excess of the yield stress. Steel can therefore redistribute load or absorb energy by yielding, which makes its behaviour ductile.

Glass can be very strong, much stronger than structural steel. However, because glass has a relatively low fracture toughness, such strengths can only be achieved when it is virtually free of defects, as a freshly drawn glass fibre might be. Normally glass will contain defects, the surface being particularly prone to getting scratched, chipped or attacked by the atmosphere. In this condition, as soon as the critical stress is exceeded, the glass will fracture. The behaviour is brittle, with no ability to redistribute load or absorb impact energy.

To counter the fragility of glass, the industry has developed a number of methods to strengthen it. These do not change its nature, but either raise the threshold at which cracking occurs, or incorporate additional ductile materials that absorb the energy and therefore prevent total failure when a unit has cracked. Among these, the three most commonly used methods are toughened or tempered glass, laminated glass and wired glass. For various reasons, toughened glass was chosen for the Serres at La Villette.

Toughened glass

Glass is toughened by heating it to about 700°C, then carefully cooling it using air blowers to cool the surfaces rapidly while the centre is still viscous. The surfaces harden and are then put into compression as the centre cools and contracts. With no overall load applied to this glass, the toughening process produces a balanced stress condition across the thickness of the glass. The high surface compressive stresses are equilibrated by smaller tensile stresses acting over a greater proportion of the cross-sectional area.

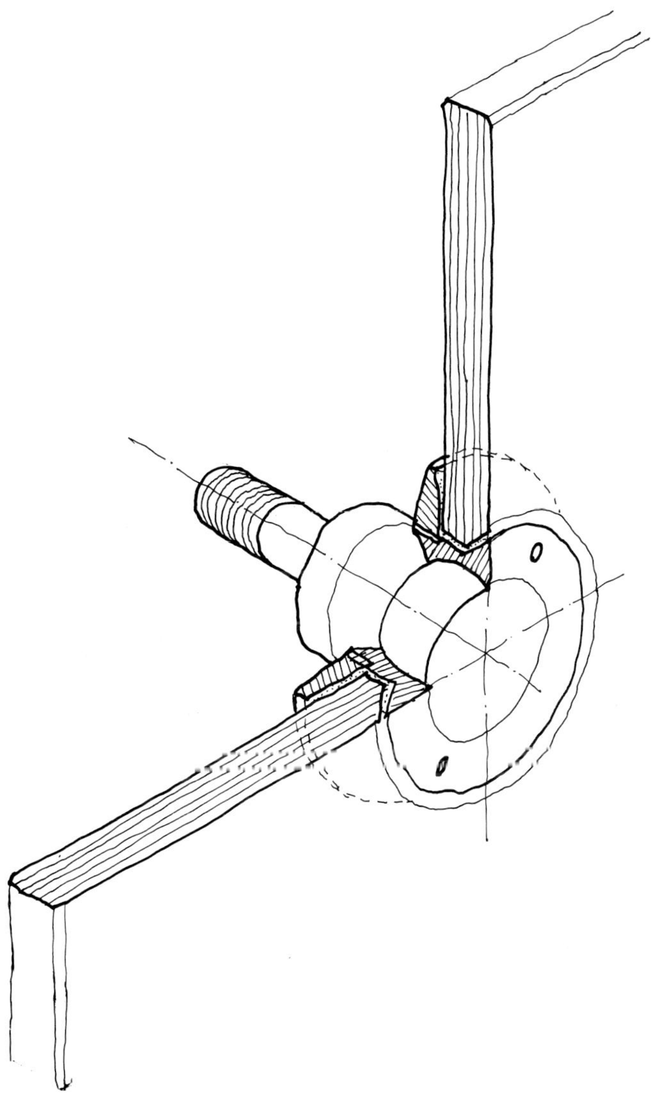

The suspension of the glass in its own plane is made possible by the articulated bolt specially developed for the Serres.

Section through a sheet of tempered glass: the external layers are in compression, the internal layer is in tension.

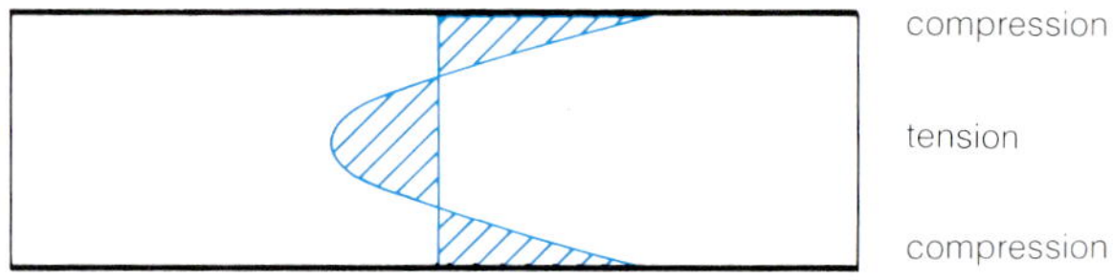

Glass failure is most likely to be initiated by surface cracks, because these tend to have the worst geometries and are subject to the highest stresses due to bending.

If the loads to which the glass is subjected do not create enough surface tension to overcome the surface compression, no crack will propagate. Toughening, therefore, increases the effective strength and impact resistance of the glass. Should an external load 'overcome' the precompression and cause a crack to propagate, then the stored energy due to prestress will cause the cracks to spread immediately in all directions and the pane of glass will fragment explosively (as was the case with the older type of car windscreens).

There is one further advantage of toughened glass where the glass is under significant permanent load – which is the case with suspended glass. In a non-toughened condition, the glass will creep and lose its strength under a permanent load. The long-term (one year or more) strength of glass under permanent load is about one-third of its initial short-term strength. This is due to the tendency of defects to increase in size.

Panes of non-toughened glass subjected to a permanent load are therefore likely to suffer instantaneous failure. By contrast, when the glass is toughened, the outer layers are always in compression and the tendency for crack growth of the material is thus counteracted.

Since it had been decided to suspend the glazing of the Serres, toughened glass was the only possible choice. This meant that all hole drilling and other machining of the glass had to be done before the toughening was carried out: any drilling or cutting done after toughening would have broken the stress relationship between the inner layer and the outer layers and would have precipitated cracking and fragmentation of the glass.

The other methods of glass strengthening were not appropriate to the Serres since none of them prevents crack propagation. Although laminated and wired

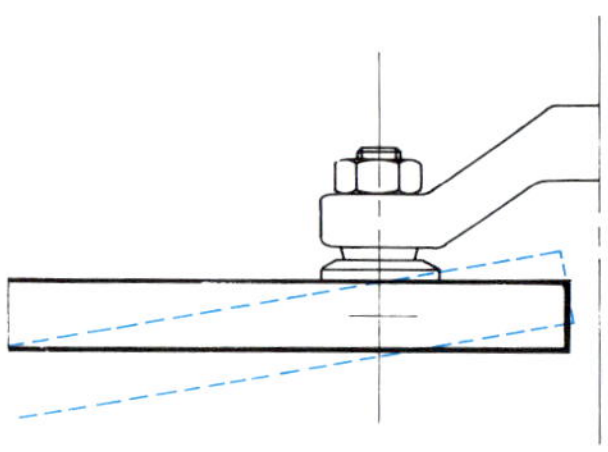

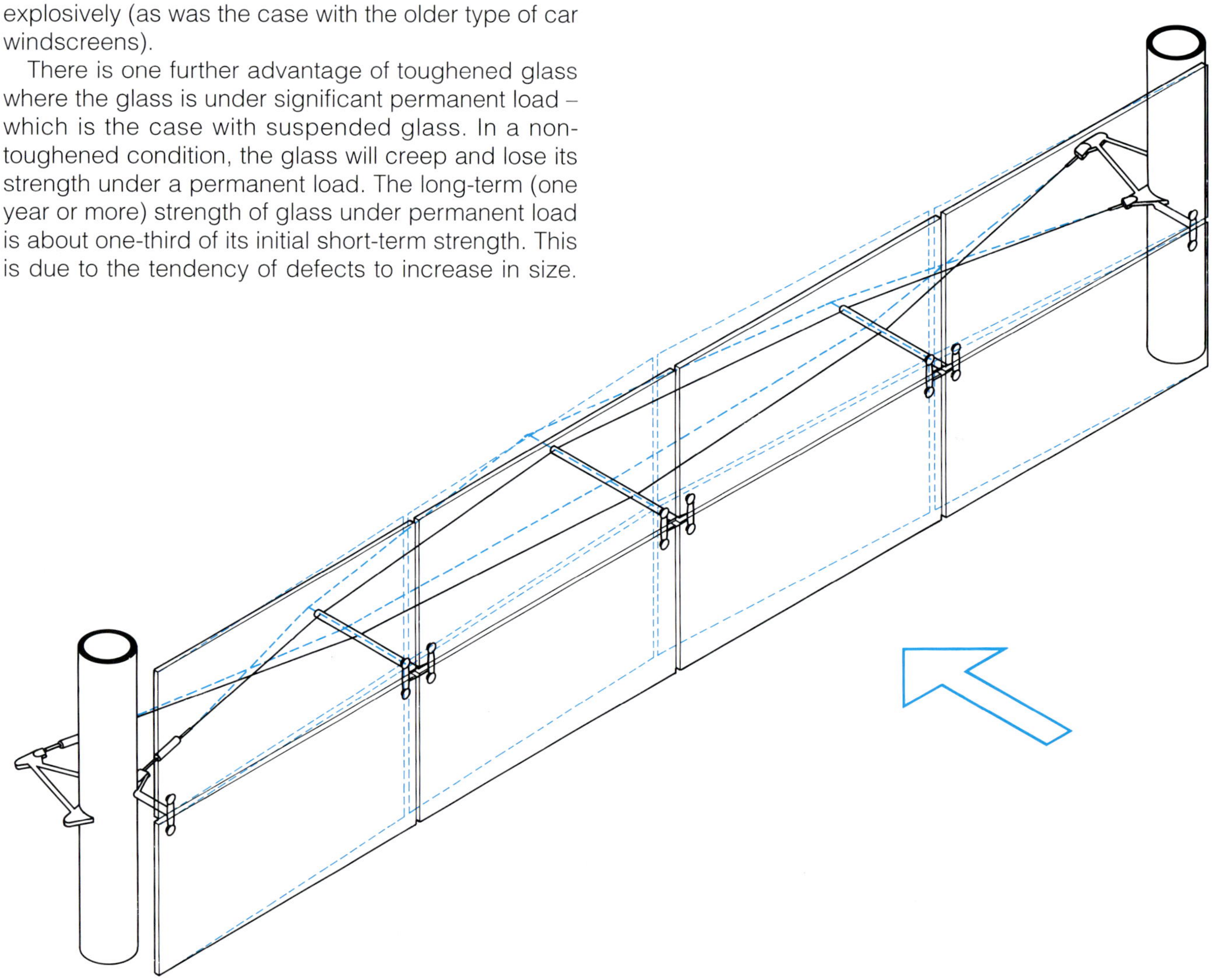

Above:
Deformations due to major loads can cause a change in angle between the glass and its suspension points.
Right:
Under extreme loads (in this case the wind), the cable trusses can undergo a deflection of 40 mm.

glass do not hold together when broken, they are no less susceptible to cracking.

Many factors can cause a variation in the level of prestress in toughened glass: the oven temperature, the length of time in the oven, the cooling time. The toughness of the glass (determined by the level of toughening) is measured with an instrument called a refractometer, which measures the change in refractive index caused by the prestress in the glass.

There are two basic types of glass toughening ovens, one in which the glass is toughened vertically and one in which it is toughened horizontally.

In a vertical toughening oven, the glass is hung from a row of tongs holding it from its top edge. As the glass softens during the heating process the tongs leave small marks on it and it elongates a little due to its own weight. This causes some slight inaccuracies in the shape of the glass sheet: the holes drilled previously sag downwards and the two vertical edges curve slightly inwards. These dimensional inaccuracies create problems in tight tolerance glass assemblies such as those at La Villette.

A horizontal toughening oven, which is more modern, can virtually eliminate these problems: the glass is fed through the oven on a series of rolling supports. The dimensions of the toughening oven were a major factor in the selection of the size of the glass panels at La Villette. The maximum width, which varies between manufacturers, is generally around 2 m. Therefore all the glass panels of the Serres measure 2 × 2 m.

Sensitivity of the support points: dynamics of the prestress structure

The cable truss system of the Serres is described in detail in Chapter 3. It will be observed that it has a considerable strength but that it suffers major distortions when subjected to extreme loads. These distortions (and the resulting geometrical adjustments) are necessary in order to achieve its ultimate resistance. The glass support components must therefore absorb these movements while avoiding a concentration of stresses on the holes. There may be a deflection of up to 40 mm in the cable truss under extreme load conditions, and such distortions cause the angle between the glass and its support points to change. Should the glass resist these movements, the glazing around the holes will be subjected to bending loads.

The hole is in fact the most sensitive part of a suspended glass structure because it is where the glass and its suspension system meet. All the loads as well as the glass's own weight and external loads (wind, impact, etc.) concentrate on the area surrounding the holes.

Taking the specific nature of glass into account, the hole must be designed in fine detail. The way in which the hole is drilled and the load applied become primary factors in the structural capacity of the suspended glass assembly. Here one must take optimum advantage of the high strength of glass, while not forgetting its fragility and sensitivity to the slightest imperfections.

Current glass support systems

To appreciate these difficulties as well as the consequences of the countersunk hole design, it may be useful to describe the behaviour of the hole according to the various structural support systems currently available. The list of five mechanical glass support systems below is followed by a detailed description of the system finally selected for the Serres.

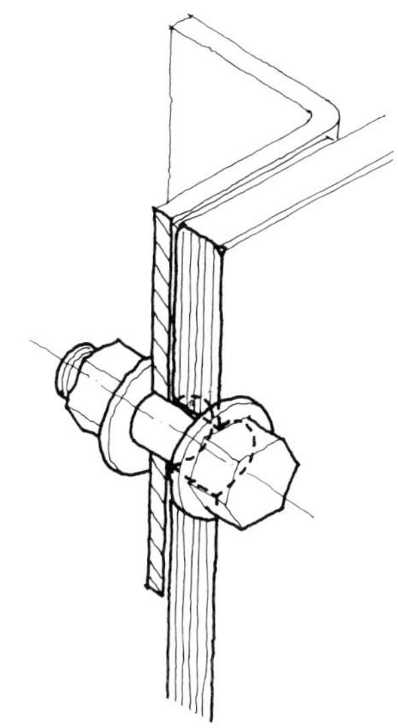

Standard bolt: The weight of the glass is taken by the area around the hole.

1. Standard bolt

The weight of the glass is concentrated at the holes in it. The glass is fixed firmly to the support structure and the assembly does not allow any differential movement between the glass and the supporting structure.

2. Patch plate

This component was used by Norman Foster in the glass wall of the Willis Faber & Dumas building. It is an improvement on the standard bolt in that the weight of the glass is not taken only by the area surrounding the hole. The patch plate is glued to the glass and the entire assembly is clamped together by a bolt. The weight of the glass is thus taken by its bonding with the patch plate, which is itself retained by a bolt. The hole (with its small drilling imperfections) is no longer subjected to the load of the glass. As in the previous system, the glass is firmly fixed, permitting no differential movement.

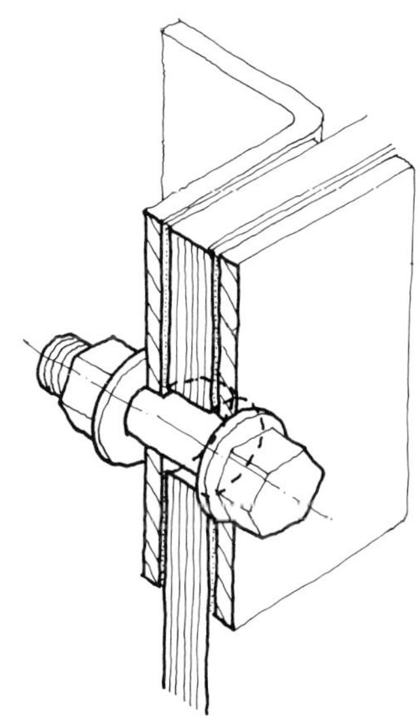

Patch plate: The weight of the glass is taken by bonding and friction against the patch plate.

3. Simple countersunk bolt

With a countersunk bolt it is possible to achieve a flush outer surface (for this reason it is commonly used in all-glass squash courts).

All the loads applied to the hole (weight of the glass, wind, impacts, etc.) are concentrated on the small

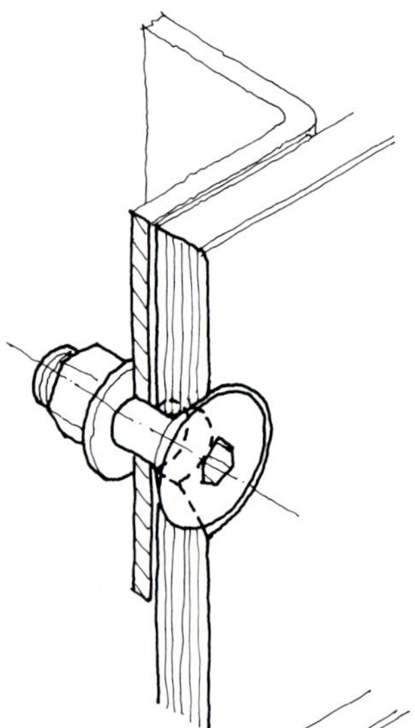

Simple countersunk bolt: The weight of the glass and the loads are concentrated around the countersunk hole.

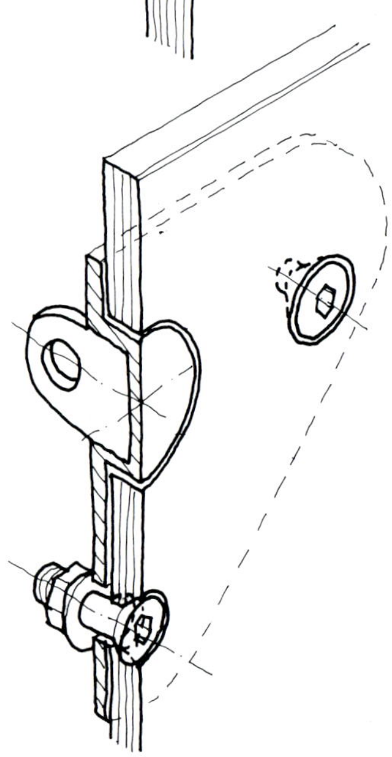

Stud assembly: The weight of the glass is taken by the stud. The other loads are taken by the areas around the countersunk holes.

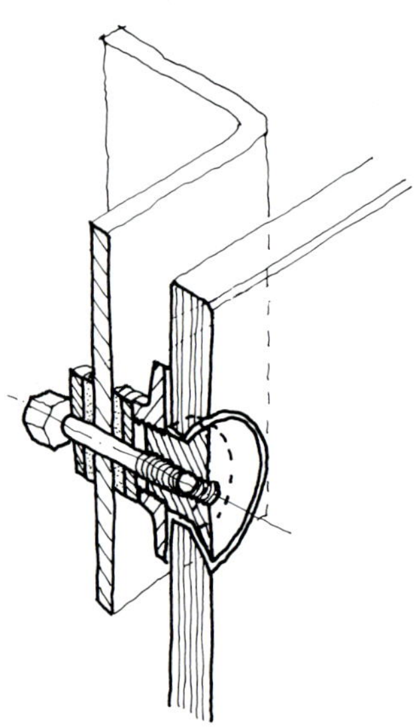

The Pilkington Planar system: Flexible washers placed at the contact points with the supporting structure allow the bolt to move in relation to the support.

area around the countersunk hole. Given the brittle nature of glass, high concentrations of stress at local points or in holes with drilling inaccuracies can easily lead to fracture of the glass. As with the other systems mentioned above, this assembly does not allow differential movement.

4. Stud assembly

This component was researched during the development phase of the Serres design; it can take relatively heavy loads while allowing the outer surface to retain a perfectly smooth appearance. The weight of the glass is taken by a large cylindrical stud and the horizontal external loads are taken by small countersunk screws. These screws fit in an oversized hole allowing them to slide vertically, thus ensuring that they will not take any vertical load. Within a suspended glass system, the weight is taken by the upper part of the hole. The larger the hole, the wider the load-bearing surface, and the pressure applied to the glass is therefore reduced. Since the support plate, stud and screw assembly do not allow any differential movement, this component was coupled with an articulated assembly in order to achieve the required movement capacity.

5. The Pilkington Planar system

This system, invented by Pilkington Glass, is widely available commercially. It allows the panes of glass to move in relation to their supporting structure while maintaining a perfectly smooth outer surface appearance. When large panes of glass bend under wind loads, a high load concentration will occur in the area of the hole if the retaining bolt is firmly fixed to the supporting structure. The Pilkington Planar system allows the bending pane of glass and the bolt to move by using flexible washers at the fixing to the supporting structure.

The above analytical comparison underlines the major design issues concerning the hole in the glass. In brief:

- a supporting system at point locations causes high load concentrations;
- the hole in the glass is very sensitive, particularly if it is countersunk;
- enlarging the hole can reduce the load in the bearing surface;
- articulated assemblies allow the differential movements between the glass and the structure to be absorbed.

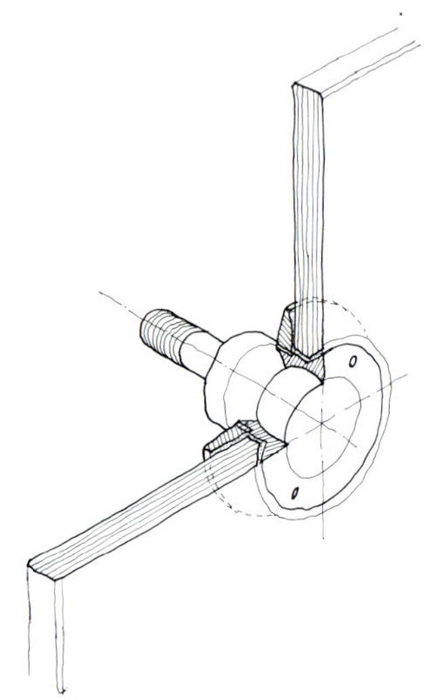

Articulated bolt: No bending or twisting moments are taken by the glass.

Articulated bolts, a system ensuring predictability

If an articulated assembly is mounted outside the glass plane, bending or twisting loads will be applied to the glass. This is the case with a stud assembly and with the Pilkington Planar system. With the articulated bolt, the system finally chosen for the Serres, the articulated assembly is positioned in the plane of the glass: this method ensures that no bending or twisting loads are taken by the glass. The diagram below, right shows a comparison between the two possible positions of the articulated assembly, i.e. outside or inside the plane of the glass. This has been confirmed by tests, which have shown the benefits that may result from preventing bending loads in the glass. The shear bolt assembly was compared with the articulated bolt system. The mean capacities were respectively 2 and 4.5 tonnes. The articulated bolt ensures that no bending load can be applied to the glass, allowing the designer to predict that all loads will remain in the plane or perpendicular to the plane of the glass. Unpredictable or parasitic effects likely to affect the supporting system are therefore excluded from the analysis. The designer knows the exact behaviour of the assembly.

The articulated bolt consists of a head which rotates freely on its stem. The stem ends in a ball on which the head is added using the standard spherical bearing techniques of the mechanical and aviation industries. The head is then machined to the required specifications: countersunk flange, thread, key holes, etc. A threaded washer is screwed onto the head and holds it against the glass. The washer is tightened to a precise torque so as to avoid any unknown or unpredictable stress which may otherwise result from the site assembly of the parts.

The liner

Steel must never come into direct contact with glass as it has too hard a bearing surface. For this reason a thermoplastic spacer is commonly used between the steel bolt and the hole in the glass. The slightly ductile nature of the liner ensures that the entire bearing surface of the hole is put to work, not just the high points. However, thermoplastics have a tendency to creep which could restrict the fixing point's capacity, eventually allowing the steel to touch the glass. Pure aluminium is used instead because it is a fairly soft metal. This choice is the result of the studies conducted by Boussois in cooperation with the central department

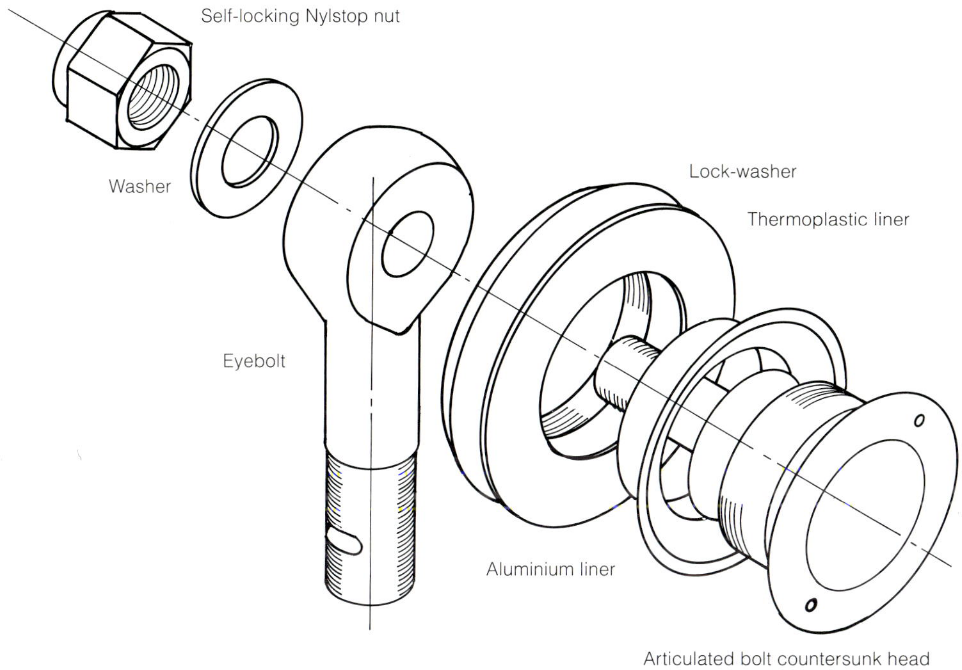

Above:
Component parts of the articulated bolt.

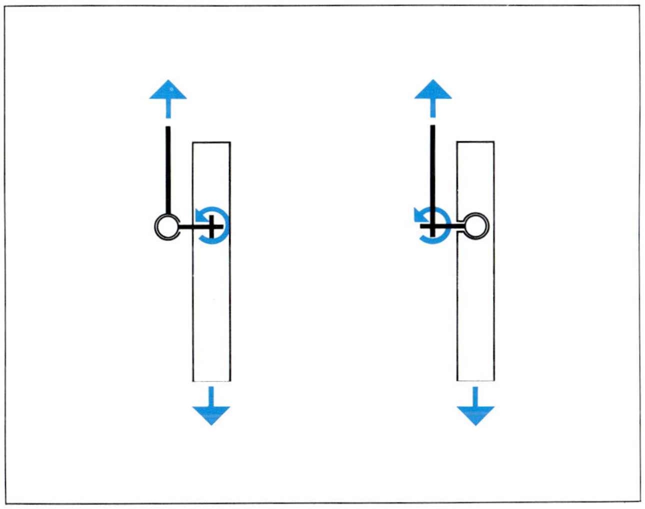

Left:
- *left, in the Pilkington Planar system or in the stud assembly, bending forces can be created in the glass;*
- *right, in the articulated bolt system, the bending forces are carried into the support.*

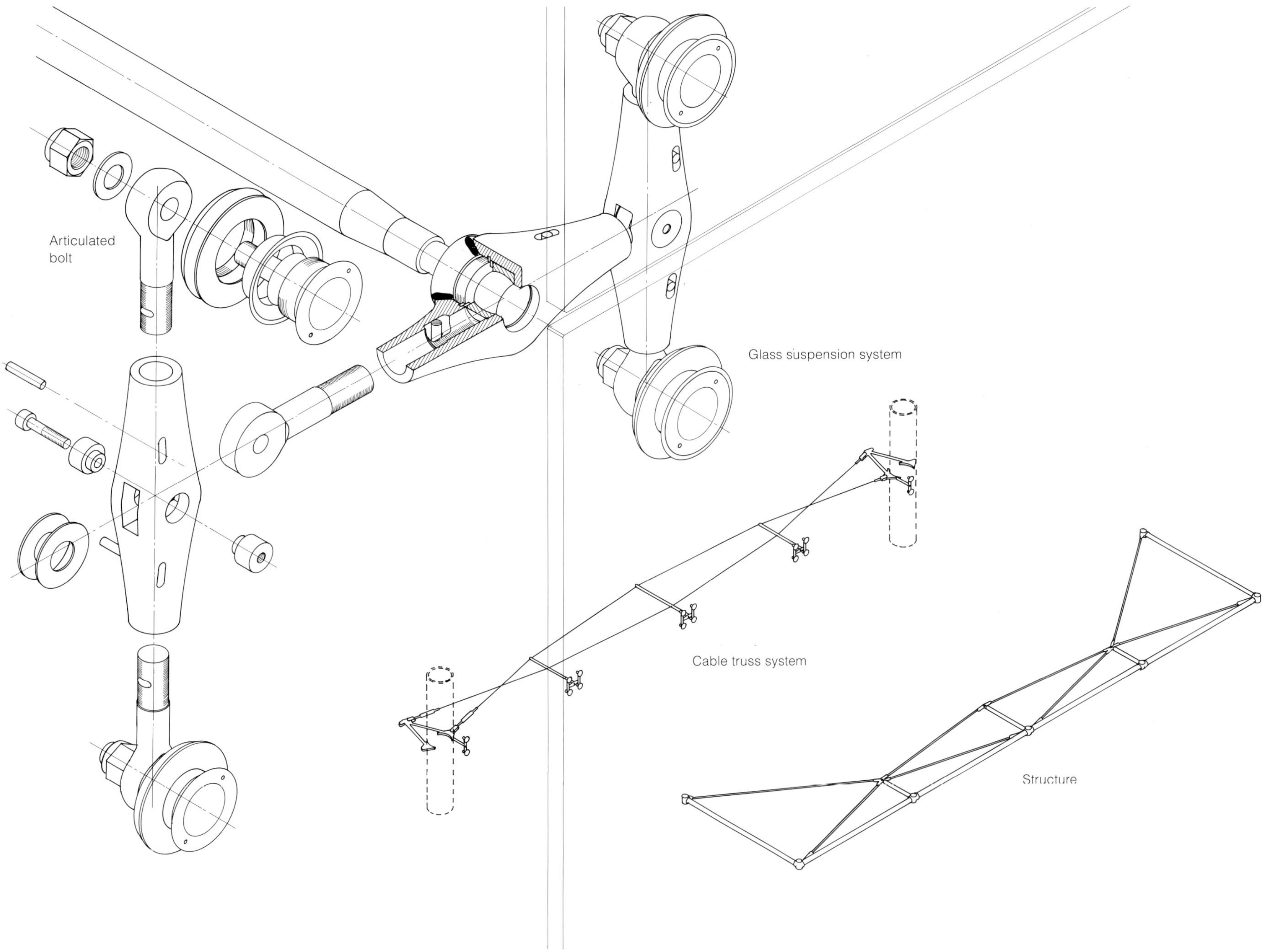
Articulated
bolt
Glass suspension system
Cable truss system
Structure

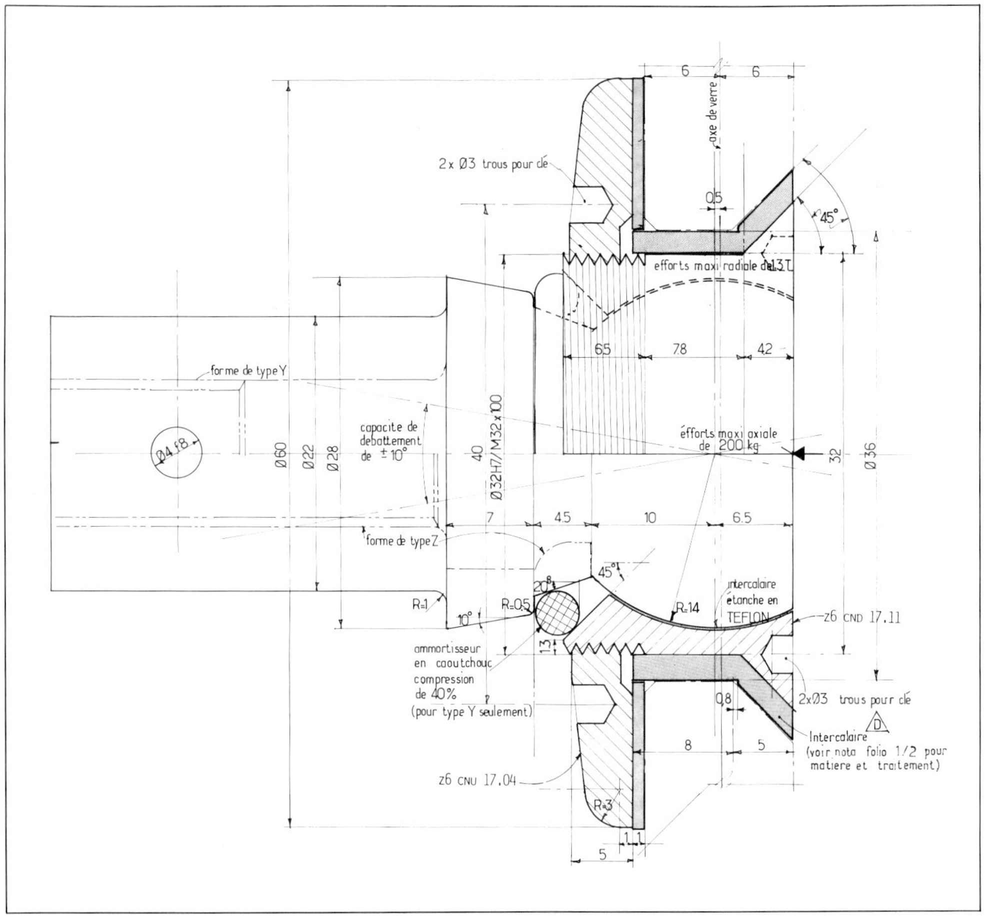

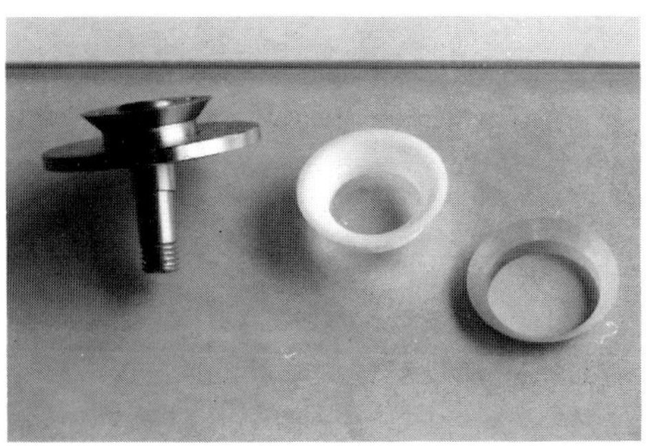

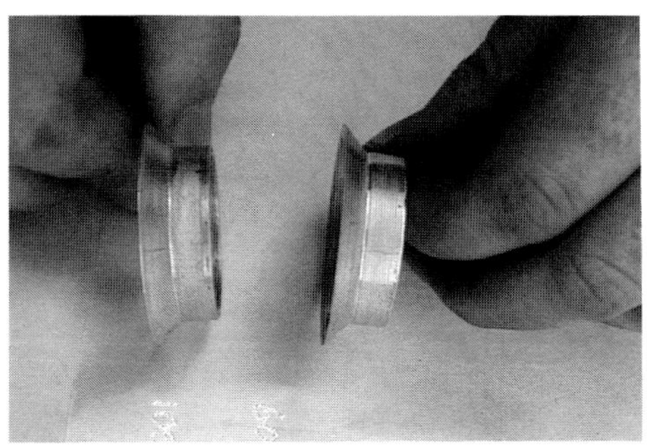

Top:
Working drawing of the articulated bolt.

Above:
Left, view of the articulated bolt with its component parts. Right, trying the liner: the imprint left by the hole in the glass varies according to the aluminium alloy tested.

for glass research of PPG (Pittsburgh, USA). Metals creep much less than plastics, and aluminium is soft enough to take the imprint of the glass and of all drilling imperfections, but strong enough to prevent any long-term deformation. The liner is finished with a Teflon coating to prevent bi-metallic corrosion between the stainless steel bolt and the aluminium liner.

Industrial process and design

Resolving the technical problems relating to the hole in the glass and to the articulated bolt requires a thorough understanding of the design, manufacture and assembly of the components used. Understanding these different stages means that they can each benefit from the other; for instance, if the manufacturing processes are well understood, this knowledge can be exploited at the design stage.

Contacts with industry

A close working relationship with industry is a key element at the design stage and a must even before the concept is determined. For instance, it was necessary to consult the industry in order to ascertain what glass sizes were feasible, even before the very first architectural drawing of the Serres was done. Similarly, the positioning of the holes in the glass, the glass thickness, the methods available for the manufacture of assembly components, etc. are all matters that are solved through consultation.

Once the concept is determined, its feasibility must be demonstrated by the manufacturers, then confirmed by sample testing.

Testing

Feedback from testing allows the design to evolve. One concept will be retained, another rejected. The limits of a design are empirically quantified, and in this way confidence in the reliability of unconventional and untried solutions can be established. Such tests have confirmed that the articulated bolt performance is superior to that of the shear stud assembly. The selection of the articulated bolt is a perfect illustration of the impact that testing may have on the project as a whole.

Manufacturing

The tests also showed the correlation between the quality of the glass drilling and the performance of the

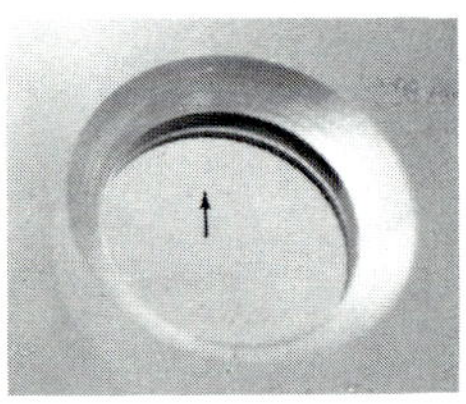

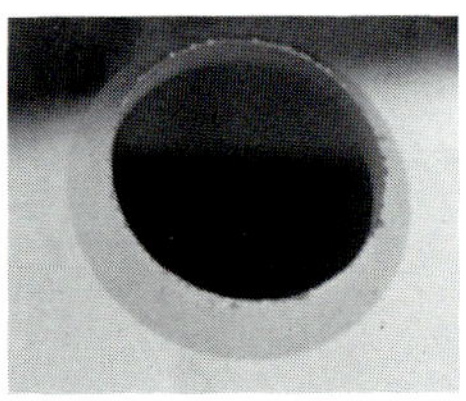

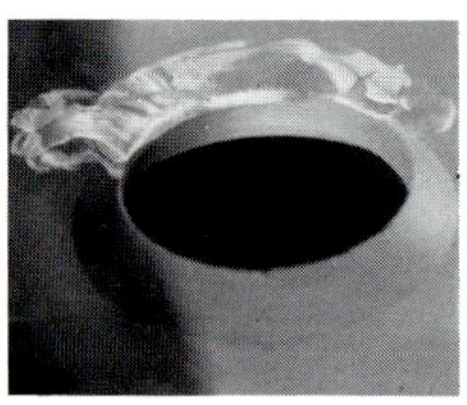

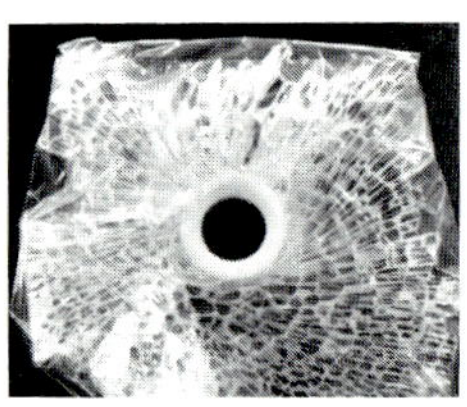

Top:
Testing the glass in tension:
Above, from left to right, and top to bottom:
Glass drilling faults and their consequences:
- *shoulder created by non-coaxial drilling from either side;*
- *spalling round the hole;*
- *worsening of the spalling following loading;*
- *complete cracking of the glass around the hole. The pattern of breaks follows the load path resulting from the forces in the glass at the time of rupture.*

assembly. The glass must be drilled simultaneously from both sides in order to prevent spalling on the side opposite the drill. If the drill bits are not perfectly coaxial, the hole will have a shoulder. Such a shoulder will cause a higher stress concentration than a smooth hole, where the load is equally distributed. The diagrams show a typical drilling sequence and how the shoulder problem can be eliminated. The first and second passes of the drill are designed to meet in the countersunk area so that with the third pass or countersinking operation, any shoulder that may have formed will be eliminated.

Spalling of the drilled area may be caused by drilling imperfections due to a blunt drill bit, the presence of foreign matter in the hole, imperfections in the drilled glass or insufficient lubrication of the bit. A hole showing spalling will also cause an unequal stress distribution. Furthermore, spalling causes crack-like defects and, given the brittle nature of glass, cracks may propagate instantly under load. It is, however, extremely difficult to avoid spalling around a perfectly cylindrical hole.

Erection

Any design must take the erection procedure into account. Fastening the articulated bolts is a complex procedure which calls for controlled tightening torques. Such an operation would be unthinkable, or would at least be badly executed, if carried out on scaffolding on a windy day. The articulated bolts are therefore designed to be correctly fixed to the glass on the ground under fairly controlled conditions. Then the whole panel (with the bolt stems protruding) is hoisted up and fixed to each support in one single bolting operation.

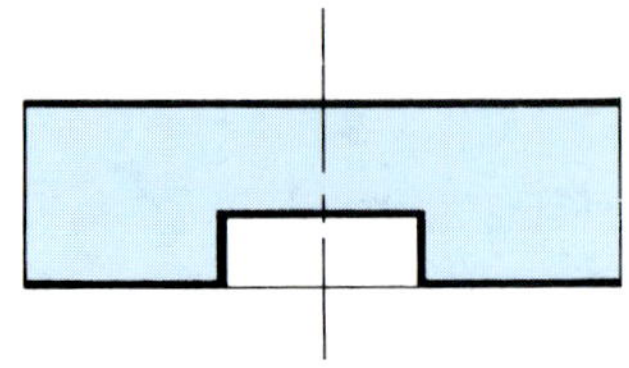

First pass

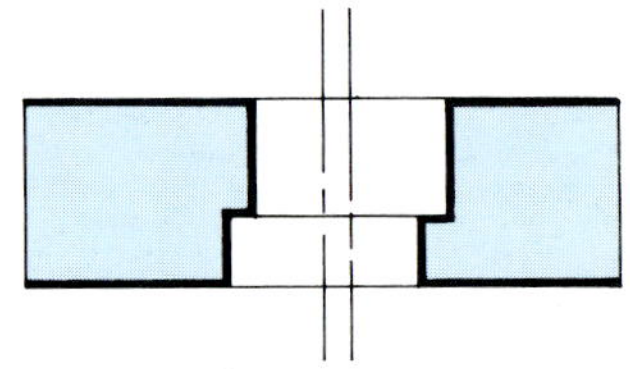

Second pass

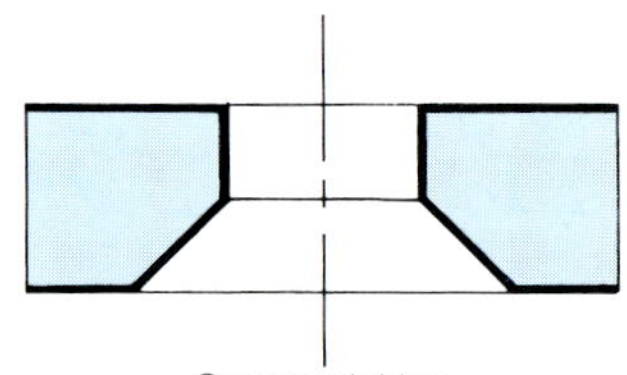

Countersinking

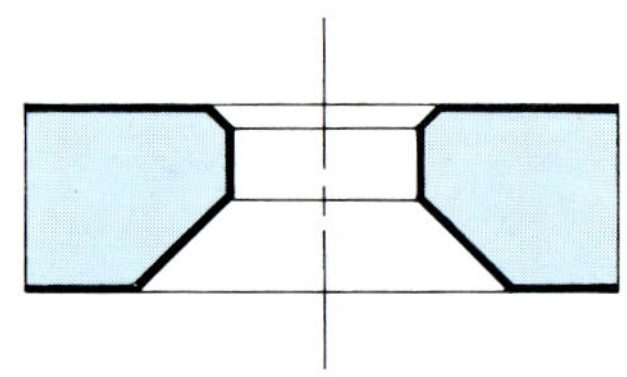

Chamfering

Top:
The stages of drilling a hole in the glass.
Above:
Drilling a hole while using a lubricating milk jet.

From left to right, and top to bottom:
On site assembly at ground level of the articulated bolt.
Below right:
After raising the glass panes the bolts are tightened onto the waiting supports of an assembly.

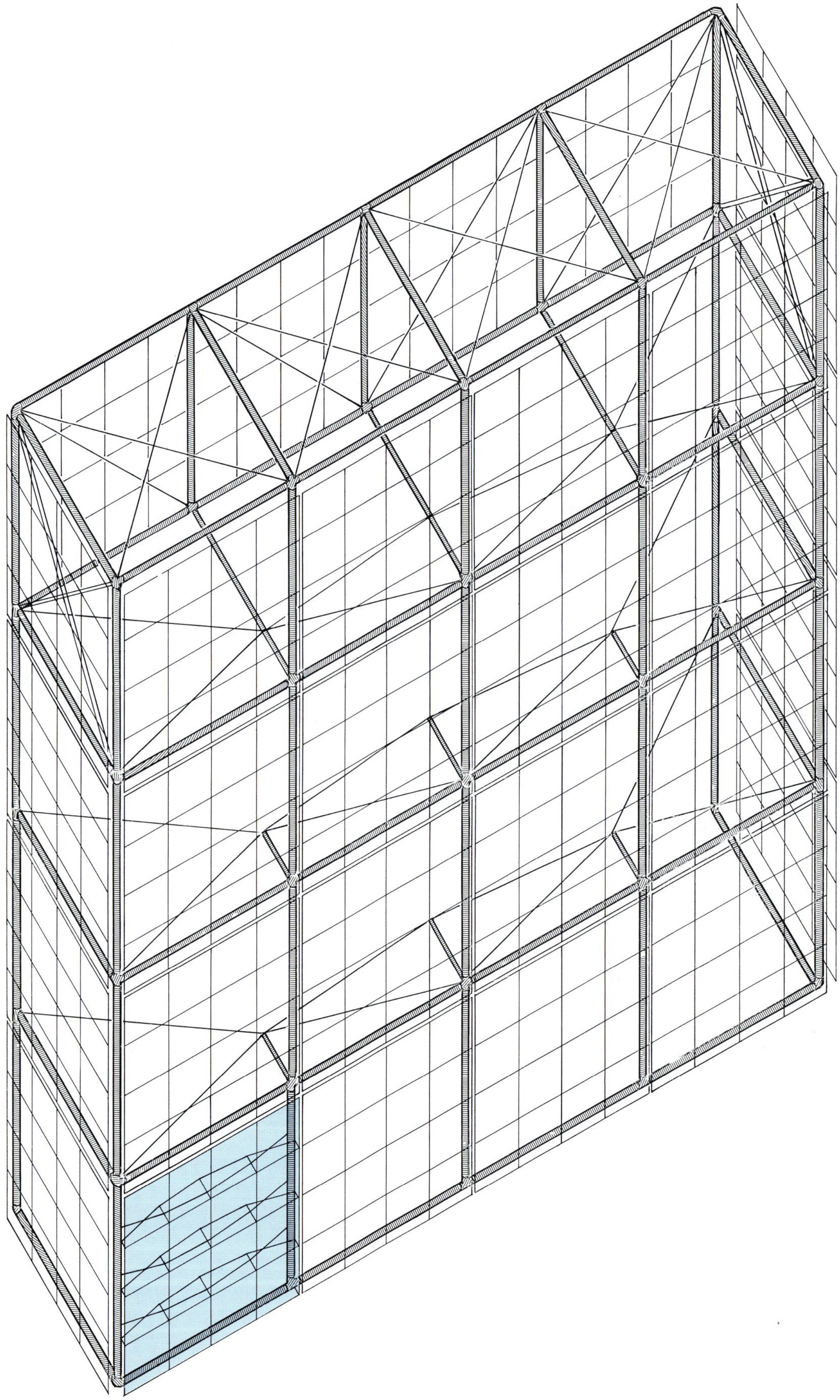

CHAPTER 2

THE GLASS SUSPENSION SYSTEM

The glass suspension system consists of a curtain of 16 sheets of glass, each 2 × 2 m, hung from the main structure with as few connection pieces as possible. The glass is hung in vertical rows, each sheet being suspended from the one above by small metal connections. This glass curtain is then braced against the wind by a cable truss system, which is described in detail in Chapter 3.

Reducing the number of connection pieces offers the obvious advantages of minimizing the interruption of the view of the park and of making the viewer aware of the plane of glass itself. The notion of transparency implies an awareness of the surface through which one looks. Aware of the plane of glass, the viewer can appreciate the fact that it hangs by itself without any other means of support. This appreciation of the glass at work is critical to the design philosophy of the project: by making the public aware of the structural role of the glass, one helps the viewer to understand the integral nature of the design and the fact that every component of the structure is interdependent.

The curtain of suspended glass

The guiding architectural and aesthetic principle which governs all details of the system is the expression of the curtain of suspended glass. The behaviour of the system must be understood in all its details and the viewer must feel that the curtain hangs freely. This is why the vertical connections between the sheets of glass are made clearly visible; the same is true of the main suspension point with its springs clearly expressed. The dynamic nature of the system is reinforced by the presence of the springs and by the appearance of all the movable assemblies.

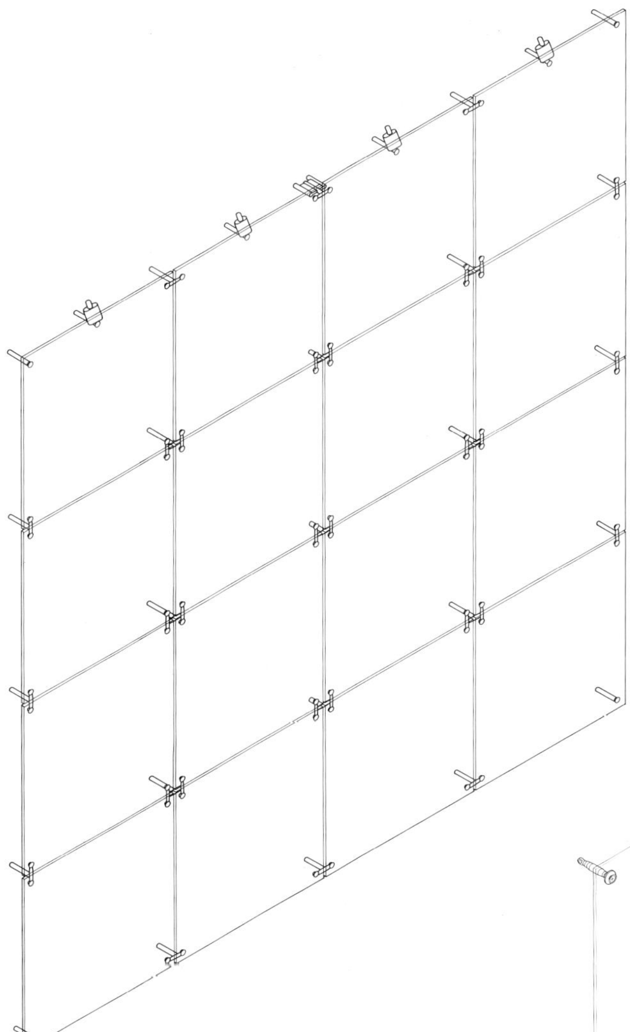

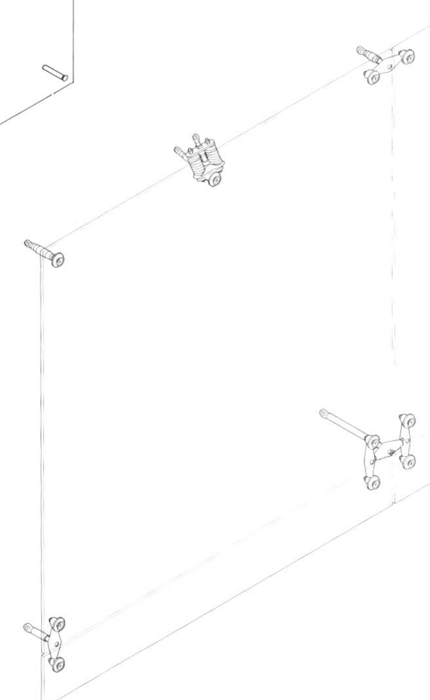

Each upper pane in one of the four columns making up a panel is suspended from its centre by a clearly identifiable piece.

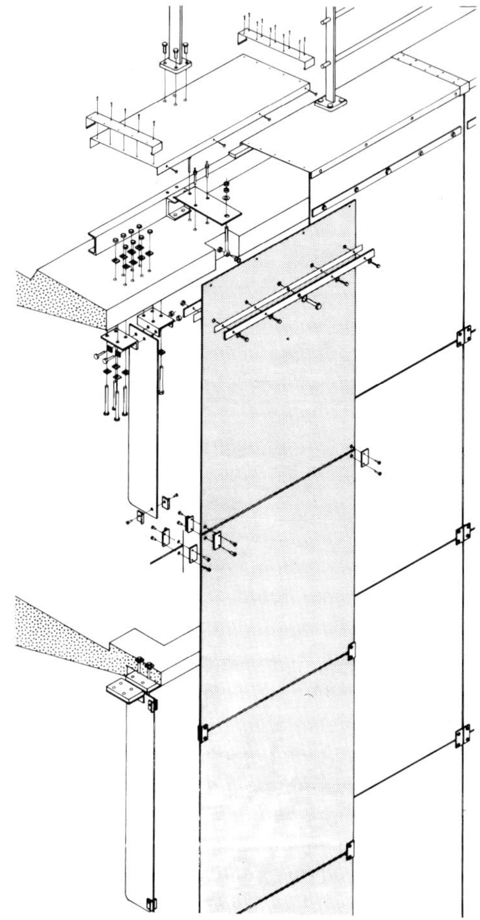

The Willis Faber & Dumas building in Ipswich (architect, Norman Foster) already applies the principle of suspended glass. According to the time of day, the curtain of glass is either totally transparent or perfectly reflective. Below left: the structure placed on the roof terrace of the first levels uses a glazed façade suspension system slightly different from that used in the curved façade.

The contribution of experience

This concept finds its origin in the suspended glass façade of Norman Foster's Willis Faber & Dumas building. The glass suspension system used on this building was considered to be an innovation at the time.

This building consists of three main levels built in reinforced concrete with an irregular shape on plan. The front elevation of the first three storeys is a continuous curtain of glass suspended from the upper floor. On the roof there is a smaller rectangular volume made of steel. It also has a suspended glass skin, but the details are treated differently.

The glass curtain of the first three storeys consists of rows of six sheets of suspended glass, each sheet hanging from the one above, with the highest one bolted to a pivoting support bar. A patch plate assembly connects one sheet to another. The curtain is then braced against the wind by glass fins bolted to each floor level using vertically sliding connections. The upper pivoting support bar and the sliding connections ensure that the six sheets of glass are always under predictable loads. Firstly, the sliding intermediate connections ensure that vertical loads external to the system can never be attracted by it. Secondly, the pivoting support bar ensures a good load distribution in the connection pieces and allows each row of glass sheets to find its own vertical equilibrium.

The suspension system of the upper unit on the roof works in the same way, but here the pivoting support bar is replaced by a suspension system with twin rocker arms. These rocker arms, used at each suspension point and at each corner of the upper glass sheet, ensure the load distribution. But in this case the suspension points are located at the two upper corners of the glass sheet and not in the centre of the top edge.

These two suspension systems – the pivoting bar and the rocker arms – were used in the early designs for La Villette. In the first preliminary design, the glass fins were suspended from the top tube of the 8 × 8 m supporting frame and fixed by a hinged connection to the lower one. This latter connection ensures that the weight of a panel can never be taken by the lower tube, while ensuring that loads perpendicular to the plane of glass will. It was therefore possible to calculate each tube for the weight of one single 8 × 8 m panel only. The same patch plate connections were then used for fixing the glass.

The Willis Faber & Dumas experience was the basis of the know-how and of the confidence necessary for designing a suspended structural glass system. The early designs for La Villette, influenced by this experience, became the starting point for the development of the current design. They offered a preliminary solution which was useful for fully understanding the problem and as a basis for improvement.

The fundamental approach of the Ipswich experience was the establishment of the concept of predictability. According to this concept, the designers can predict precisely how the various components of the design will behave. Therefore all the assemblies are designed so as to ensure that only the specific loads for which they are designed will be applied, to the exclusion of all others. The articulated bolt described in Chapter 1 is a good example of the application of this principle. At Ipswich, as in the early designs for La Villette, the articulated supports, the glass fins and the hinged connections all follow the same principle.

A flexible supporting structure

The Willis Faber & Dumas building is fundamentally different from the Serres design in that, at Ipswich, the glass-supporting structure is relatively rigid and the only structural movement that can affect the glass is the deflection of the concrete floors due to creep and live loads. At La Villette, on the other hand, three types of deflection need to be taken into account. Firstly, the Vierendeel main frame* can deform under strong wind loads. The side panels of the Serres are exposed to cross winds which, in extreme conditions, can cause the front façade to deflect into a parallelogram shape; each 8 × 8 m panel is then affected.

Secondly, the top suspension tube of each 8 × 8 m panel can deflect under the weight of snow on the roof and under live loads from the maintenance systems which are also fixed to this tubing. Thirdly the deflections of the cable truss must be taken into account (the effects of the cable truss deflections on the glass

*A Vierendeel truss (or ladder truss) is a truss that has no diagonal bracing members. The stiffness of the truss is achieved by using rigid nodes to connect the vertical and horizontal members.

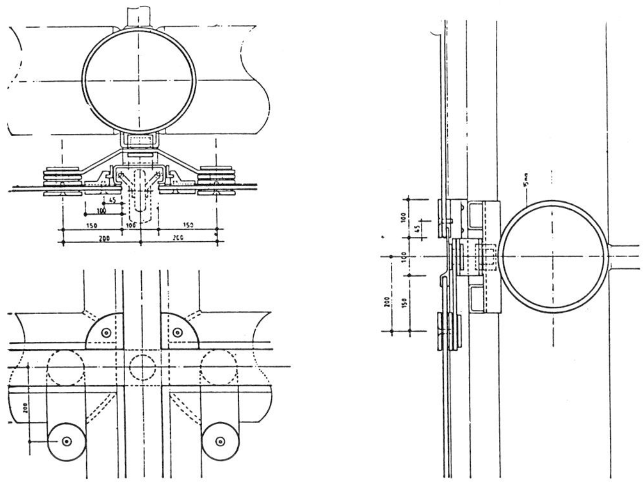

Left and below:
Initial drawings for the Serres at La Villette take up the solution (adopted at Ipswich) of glass fins.

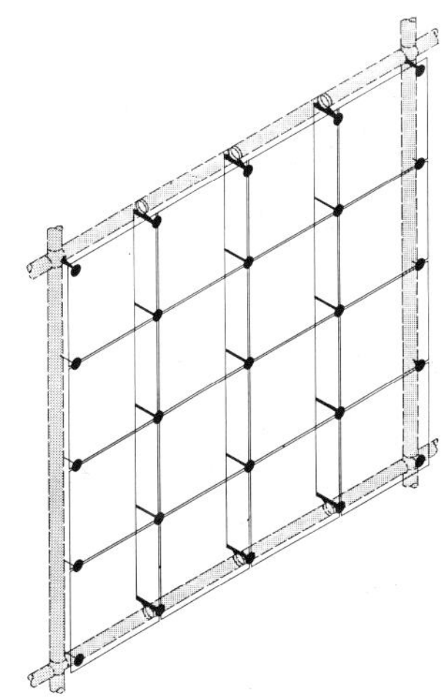

holes have been discussed in Chapter 1). Similar effects are likely to affect the glass suspension system.

The first two deflections mentioned above are particularly significant because they cause deformation of the plane of suspended glass. Each sheet of glass remaining rigid and perfectly square, any movement is thus concentrated at the support points between two sheets and between the glass curtain and its supporting structure. Deflections perpendicular to the plane of glass (deflection of the cable truss, for instance) are concentrated around the holes drilled in the glass.

Behaviour of the glass curtain

As the underlying principle of the suspension system design is predictability, the suspension elements must be such that the designer will be able to predict exactly how each panel will behave under the various deflections and movements discussed above. The performance required of the glass suspension system is very simple: on the one hand, it must take the weight of the glass (a perfectly vertical load); and on the other, it must brace it against wind loads (perfectly perpendicular to the plane of glass). It does not take any load coming from other directions; for instance, it offers no resistance to the loads caused by the main frame deformation, and thus applies no unpredictable loads to the glass.

There is however one exception to this rule: a horizontal connection located in the plane of the glass is required to position precisely each array of 4 × 4 sheets in an 8 × 8 m bay.

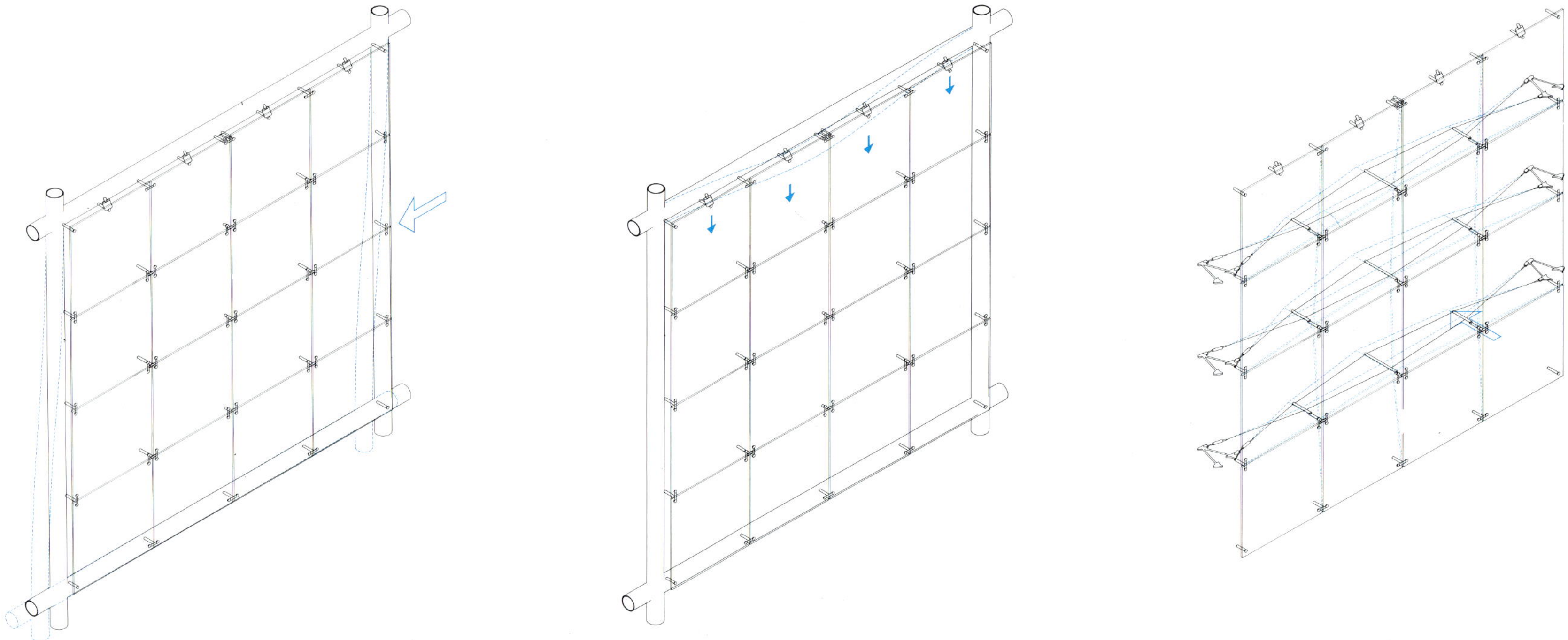

Above:
The three types of deformation which the façade can undergo. From left to right:
- *lateral deformation of the framework under the action of wind;*
- *bending of the top tube of the framework as a result of the weight of the glass and maintenance equipment;*
- *deformation of the cable trusses.*

Right:
The joints between two sheets are articulated. Lateral movement is therefore possible.

Taking the weight

The glass is suspended in vertical rows of four sheets, one above the other, each connected to the other by connection pieces at their corners. The upper sheet of each row is then hung from the main frame at the centre of its top edge. With this central suspension point, the glass is able to find its own balance and to hang perfectly vertical, irrespective of the straightness of the support tube. Furthermore, this single and central suspension point guarantees that no movement can affect the system, not even a lateral load. If the rows of four sheets of glass were fixed using two rigid suspension points, the system would then resist lateral loads and would have to be designed to take these as well as the weight of the glass.

Each sheet of glass is then individually suspended from the one directly above by two-hole connections fixing the distance between the horizontal edges of two sheets of glass so that the joints are of even width. To ensure that the connections do not attract lateral loads, they are articulated and can therefore rotate sideways.

A bay consists of four rows of four sheets of glass, each row hanging independently of the others. All horizontal connections are hinged, thus preventing all possible transfer of vertical loads – either from one row to an adjacent one or into the frame of the main structure behind the glass. The fact that the rows can slip

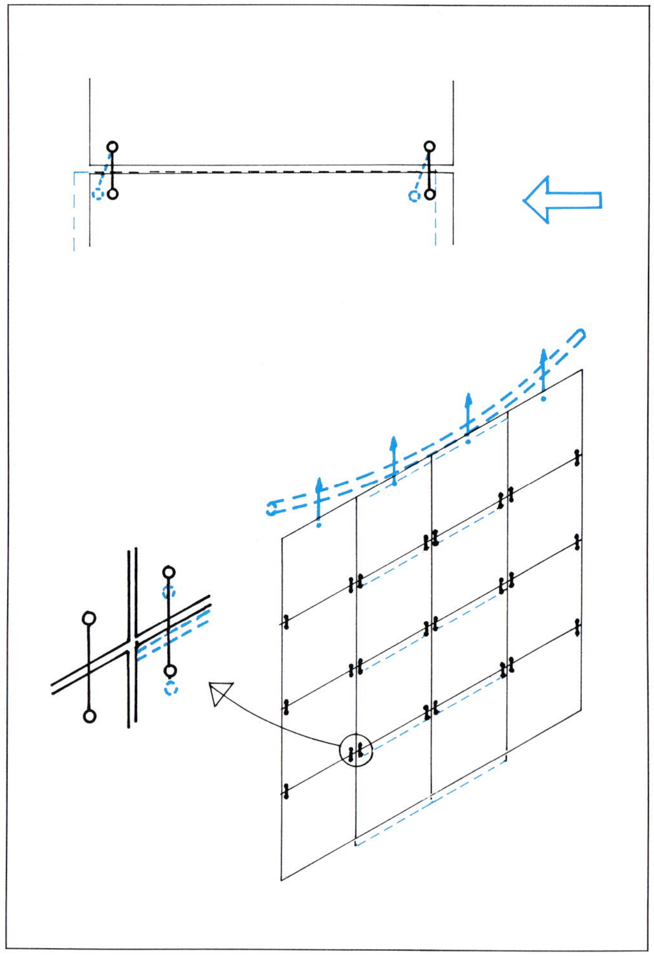

relative to each other means that the top tube can sag without any change in the hypotheses concerning the loading of the glass structure.

Resistance to the wind

Each 2 × 2 m sheet of glass is braced against wind loads at all four corners. Wind loads are transferred either into the cable trusses or into the perimeter 8 × 8 m main structural frame. At all mid-panel connections, the two vertical elements of the four-hole support assembly are linked by a cross-bar, itself connected to one cable truss strut. At the perimeter of the panel, a connecting bar fastens two sheets of glass (two-hole connection) and links them directly to the main structural frame. A small strut (one-hole connection) links the glass sheet to the main structural frame at all four corners of each panel. All these connections are hinged in two places so that they do not resist any lateral loads parallel to the plane of glass. This freedom from lateral restraint means that the plane is not affected by the tendency of the main frame to adopt a parallelogram shape under lateral wind load conditions.

Special case: horizontal positioning connection

The only part of the suspension system offering a resistance to lateral loads in the plane of the glass is in the middle of the top tube of a main structural frame bay. It is hinged vertically and so does not alter the structural hypotheses when there is deflection in the main frame top tube; nor does it interfere with the freedom of the panels with regard to the main frame parallelogram action, since it is located on the top tube itself. Temperature fluctuations cause the stainless steel Vierendeel frame and the glass to expand and contract at different rates. The fact that there is only one fixed point between these two systems in the plane of the glass and that all other connection points are free to rotate allows for differential thermal expansion.

The spring function

Although each of the four rows of a panel are theoretically free to slip relative to each other, the resistance provided by the silicone weatherproofing of the joints must be taken into account. In fact, an entire 8 × 8 m panel has a tendency to behave as one single sheet. Lateral loads in the plane of the glass would create support moments and might cause significant changes in the loads at each support point. This is why a prestressed spring mechanism is incorporated into each support bracket in such a manner as to ensure that the weight of the whole panel is always evenly distributed between all the support points. Each mechanism remains rigid until it is subjected to loads greater than the weight of the glass and fittings, which is approximately 600 kg in normal circumstances. Should any one of the springs be subjected to a load of more than 600 kg, it will sag until the

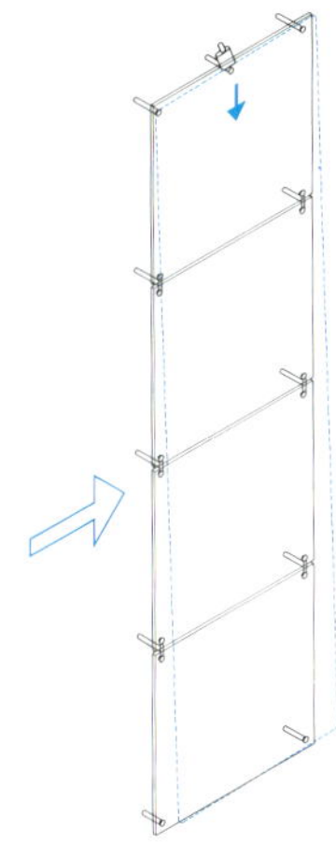

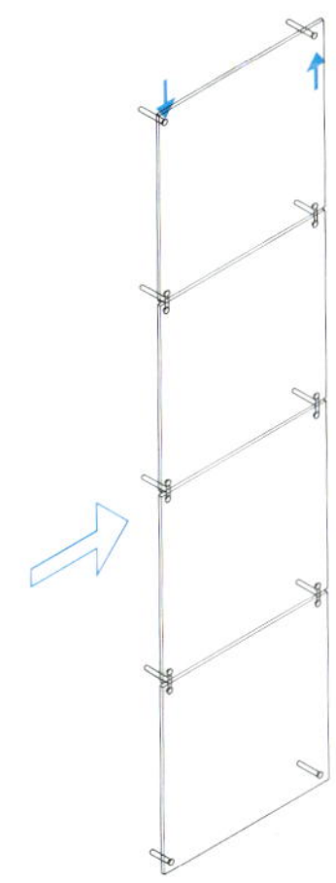

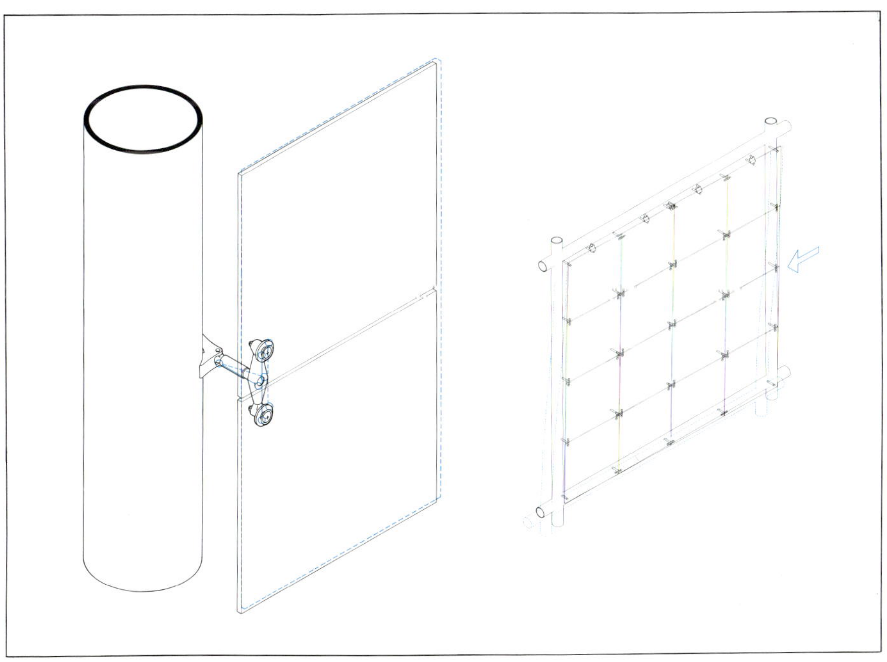

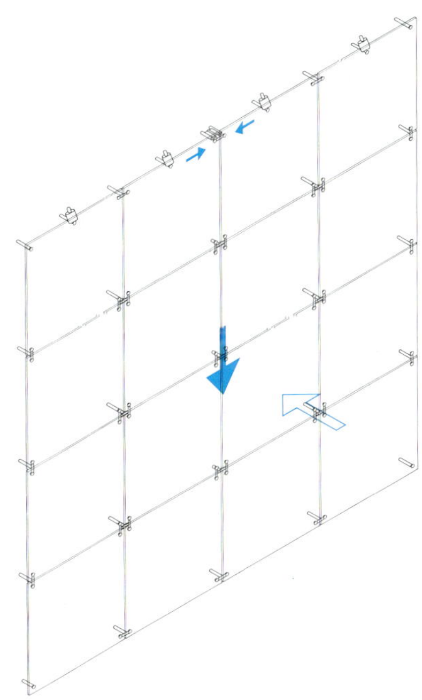

Above:
As the suspension point of a column of sheets is central, the system does not take up any lateral loads.
Far left:
At the perimeter of a panel, an articulated bar directly links the assembly to two points on the framework. The glazed plane is therefore not affected by possible deformations of the framework.
Left:
Only the assembly for horizontal positioning of the glass panes can take lateral forces.

The principles of suspension and wind-bracing of a panel.

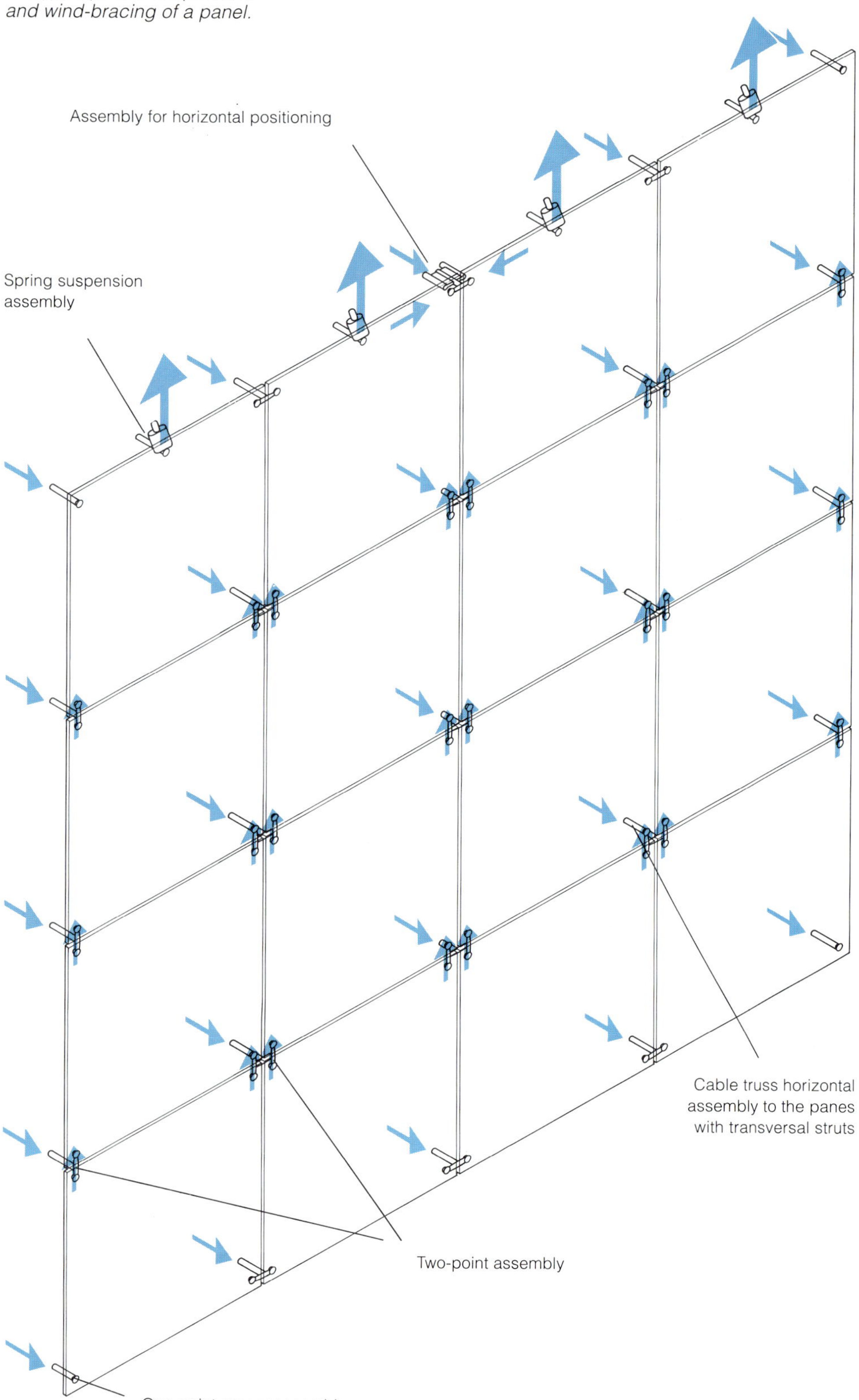

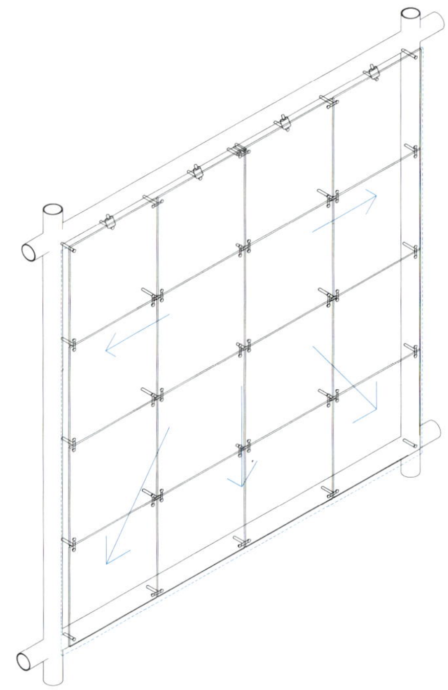

remaining support points take up the load. This 'fuse' action is necessary to enable prediction of the loads that can be applied to the glass and to each support point of the main structure frame. The springs also act as shock absorbers for the entire glass system in the event of a fracture in a sheet of glass and the consequent instant change in the load path. The diagram on this page is a summary of the concept for the support and bracing of a suspended glass curtain.

Assembly tolerances and adjustment capacity

The designer must provide capacity for adjustment to allow for the manufacturing and assembly tolerances of the system. It must be possible to adjust each element according to the differences discovered, on site, between the theoretical dimensions and the real dimensions of the structural main frame and of the holes in the glass sheets.

Small inaccuracies can be found even in the most precise work, and the accumulation of these inaccuracies can produce considerable dimensional problems at the time of installation. Some inaccuracies are unavoidable, such as the position of the holes in toughened glass: it has been seen that the glass changes shape slightly in the toughening oven as a result of the extreme heat. The glass suspension sys-

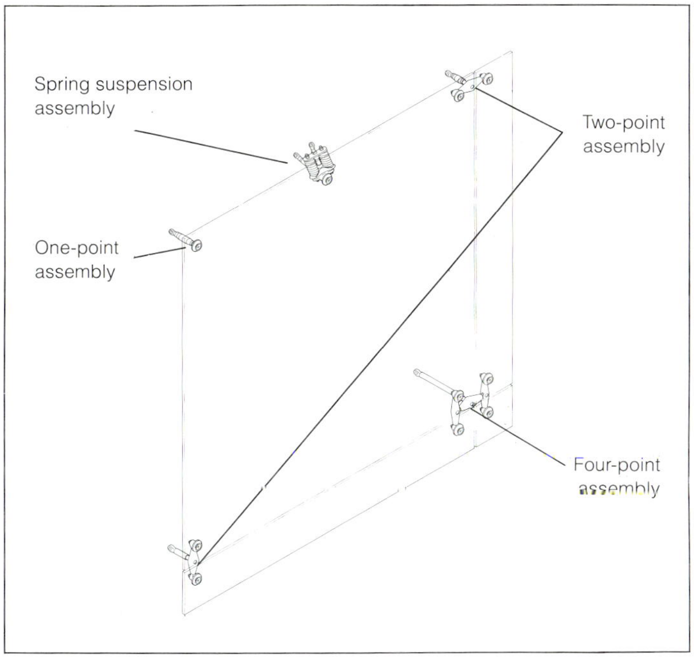

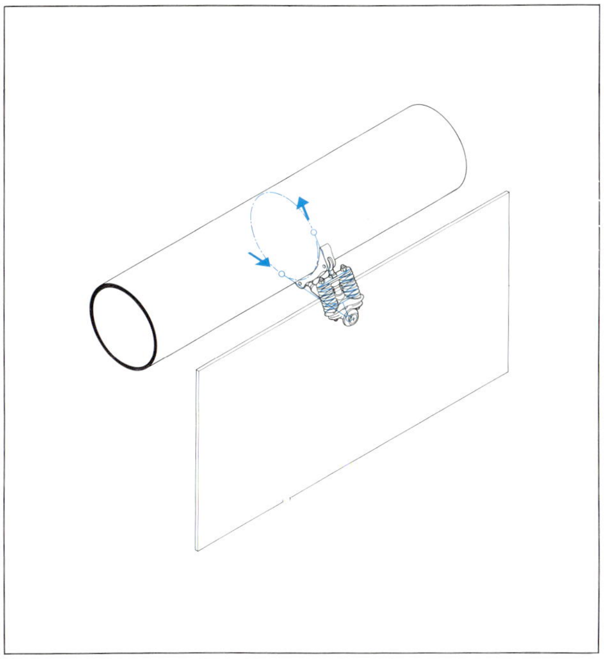

Left-hand page, right:
The springs in the suspension assembly allow an always even distribution of the forces in a panel.
This page:
Left, the four glass suspension systems.
Right, the sprung suspension assembly allows the load from the weight of the glass to be transferred into the framework.

tem consists of screw thread assemblies which enable the length of the components to be changed in order to accommodate dimensional errors. These assemblies must be locked using lock nuts or pins once the final adjustments have been made, so that they do not become accidentally unscrewed.

Assemblies can become adjustable simply because they are hinged. For instance, the spring support assembly has a double hinge which makes lateral adjustment possible. If a rigid bracket were used instead, and should it be inaccurately welded to the main structure frame, it would be impossible to correct this defect when fitting the glass sheet.

Details of the glass suspension system

The glass suspension system is made up of four different support systems: the spring support assembly on the one hand, and three types of wind bracing connections on the other, which pick up one, two or four holes drilled in the corners of the glass sheets.

The spring support assembly

The shape of this assembly is a product of its function, which is to transfer the load of the glass into the support structure across the spring 'fuse'. Its geometry is determined by the fact that the structural tubes are best loaded tangentially, so that the loads are evenly distributed around the tube section (perpendicular loads could deform the tube into an oval shape). This assembly is fastened to the horizontal tube by a cast component with two pierced lugs (from which the assembly hangs). The quadrant shape of this component allows the tangential loads to be directly transferred into the tube.

The V-geometry of the assembly divides the loads of the glass weight into an oblique tension component through the spring assembly and into a horizontal compression component through the compression strut or body part of the assembly. The glass is hung from this assembly by the articulated bolt described in Chapter 1 (p. 37 onwards).

This and all the other suspension connections consist primarily of stainless steel lost wax castings which are then machined to receive precision mechanical components such as spherical bearing eyebolts.

The freedom of shape achieved by casting the body allows a number of difficult mechanical assembly and geometry problems to be solved using a single component in the shape of a bird: the two 'wings' receive the studs from the spring assembly, the 'head' the articulated bolt and the 'tail' the spherical bearing eyebolt that connects the assembly to the support

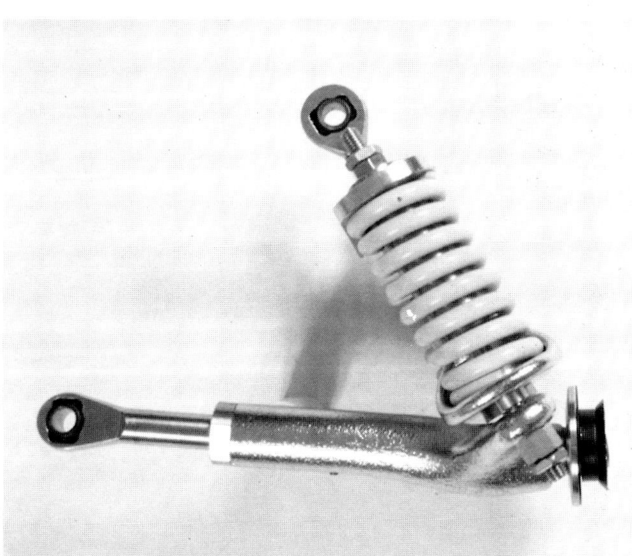

The sprung suspension assembly:
- *front view showing component elements;*
- *side view of the complete assembly.*

structure top tube. This eyebolt is mounted in a threaded sleeve which, when screwed in, alters the length of the body of the assembly for site adjustment. A similar threaded sleeve is used for the same purpose in the spring assembly.

The spring assembly consists of two helical springs compressed between two plates (for dimensional reasons two springs are necessary; a single spring would have been too big for the space available). The springs must be carefully calculated to ensure that they will achieve a specific length under a given load: 740 kg is the component of the 600 kg load taken by the spring under normal conditions, in other words the weight of the glass plus that of the attachments. The spring assembly must remain fixed under loads of up to 740 kg; but when loads above that threshold are applied, it must sag. The assembly is prestressed to this 740 kg load by using springs with a free length of 150 mm compressed to the required length of 100 mm. The prestress is taken by the threaded rods holding together the two end plates (top and bottom) of the spring assembly. The bottom plate is a lost wax casting, specifically shaped to transfer the loads from the two springs into the centre of the assembly and to prevent the springs from slipping sideways.

A cylindrical component, which acts as a sliding guide when the springs are sagging, is attached to the centre of this plate; another cylindrical component fits inside each spring and acts as a spring guide to prevent the spring from buckling sideways. These guides are also used to limit the sagging of the springs.

The spring assembly is fixed to the bird-shaped body of the casting by simple bolts and lock nuts. The 'wings' of this casting and the nuts have spherical surfaces to allow the variable V-geometry of the system to open or close when the length of each half is adjusted.

One-, two- and four-hole connections

The remaining components of the glass suspension system all belong to the same family. They connect one, two or four holes either to the perimeter structural frame or to the cable truss. In some cases, their only function is to enable these connections to resist wind loads perpendicular to the glass surface. The nine four-hole connections at the centre of each panel, and the six two-hole connections connecting the glass to the vertical tubes at the sides of the same panel, also take the weight of the glass. Finally the connection for the horizontal positioning of the glass (located in the centre of the top rail of a panel structural frame) resists

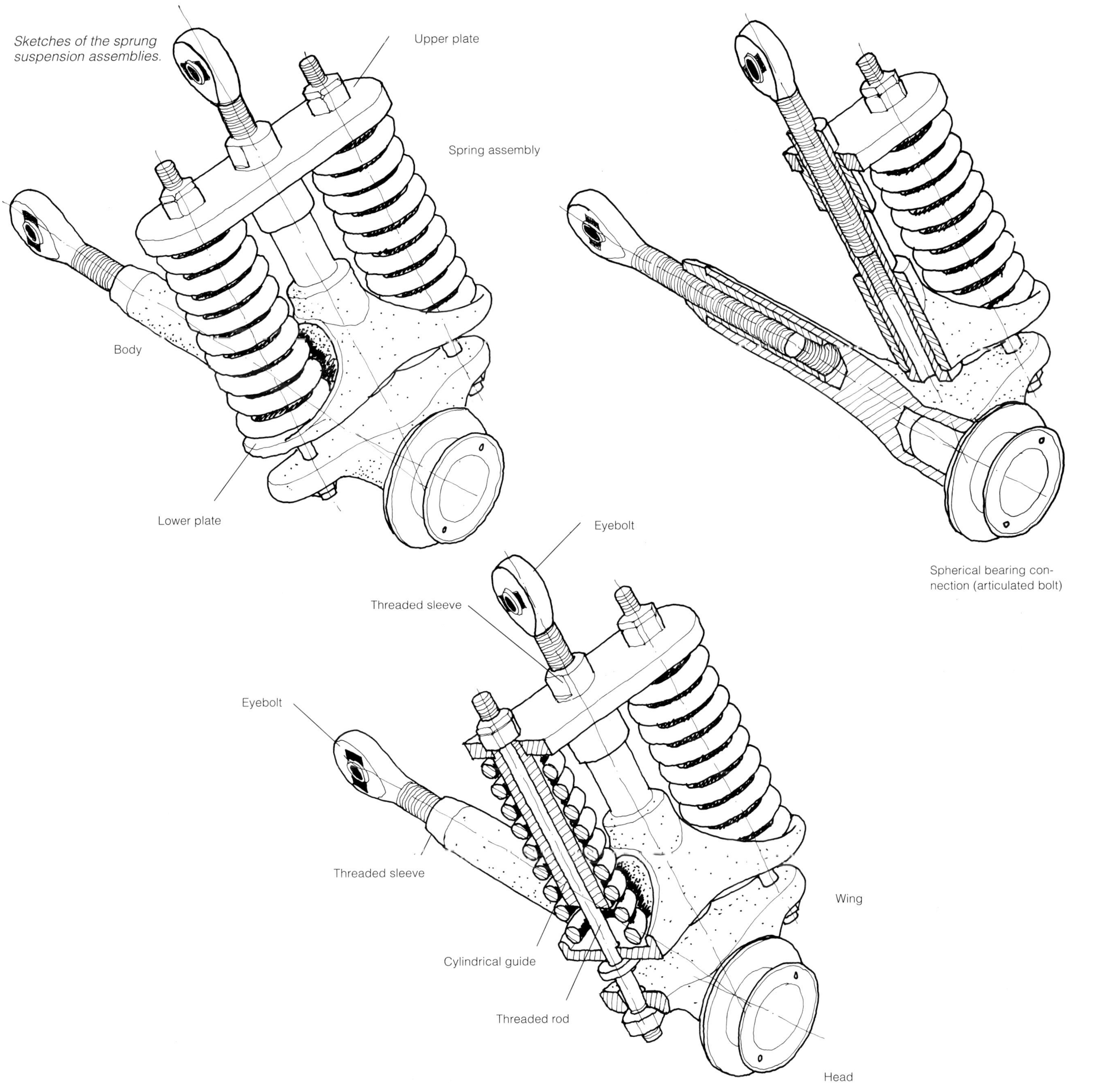

Sketches of the sprung suspension assemblies.

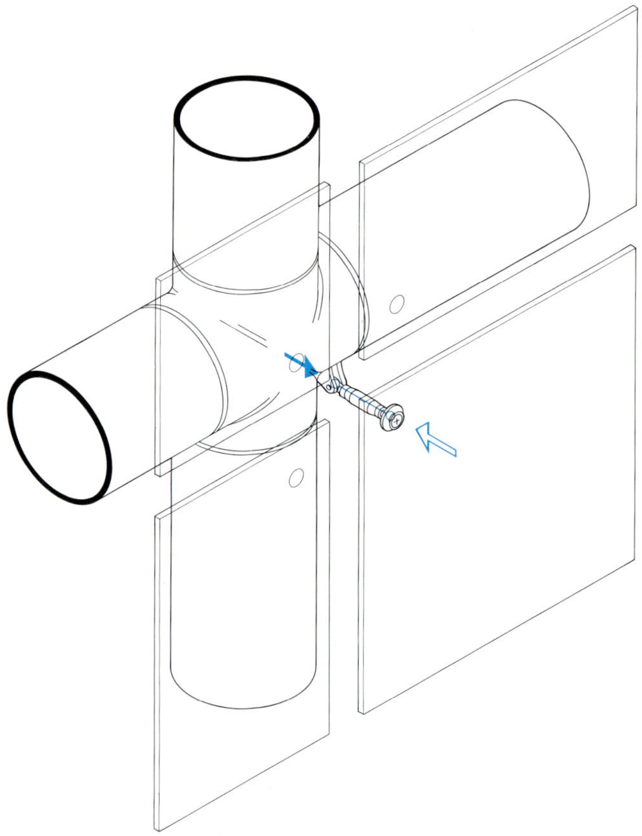

ONE-POINT ASSEMBLY
Opposes only the horizontal loads perpendicular to the glazed plane.

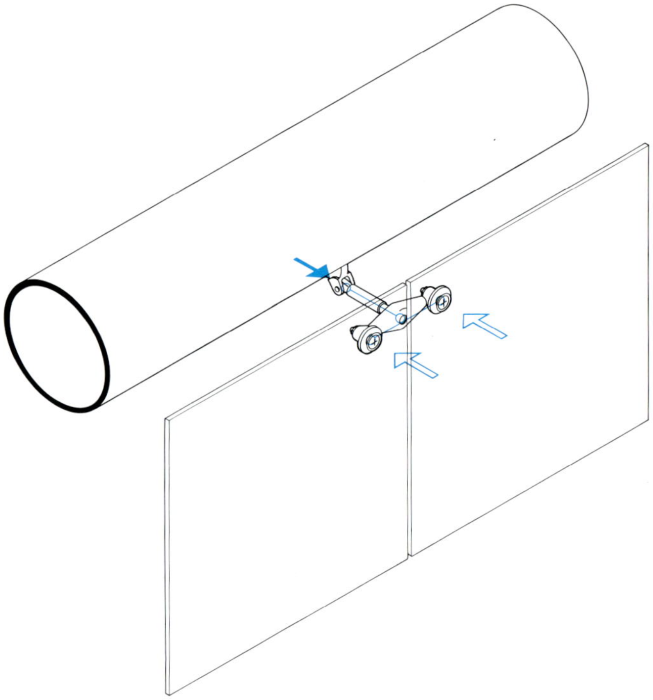

TWO-POINT ASSEMBLY
Opposes the horizontal loads perpendicular to the glazed plane.

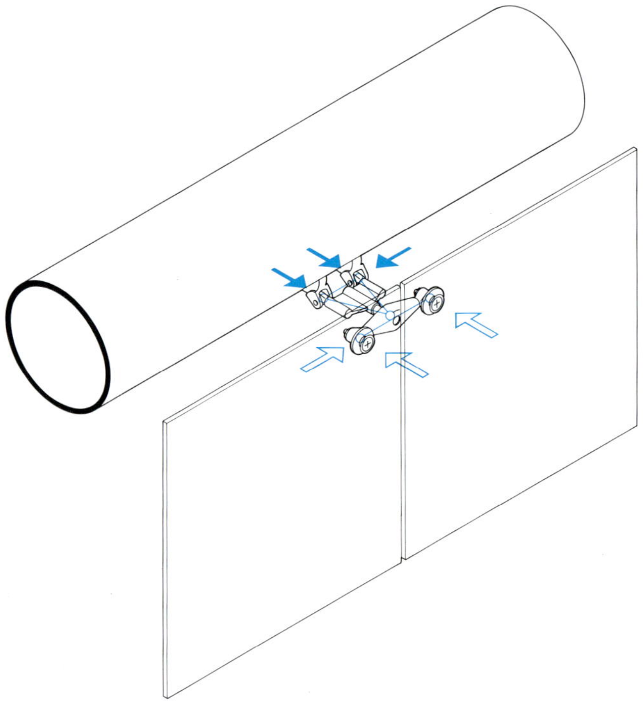

TWO-POINT ASSEMBLY FOR HORIZONTAL POSITIONING
Opposes all horizontal loads.

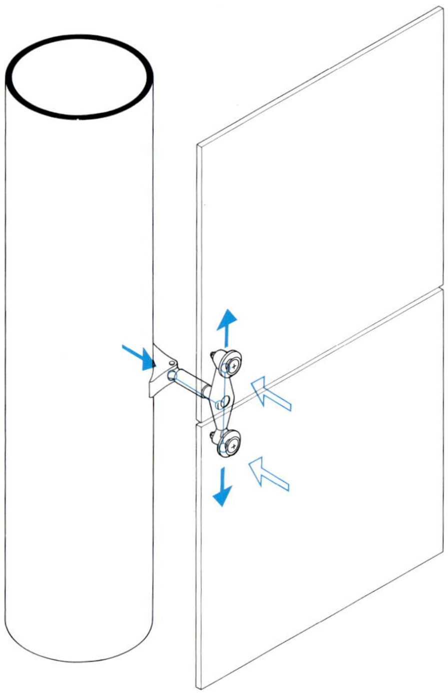

TWO-POINT VERTICAL ASSEMBLY
Opposes the horizontal loads perpendicular to the glazed plane.
The glass's own weight is transferred to the upper panes.

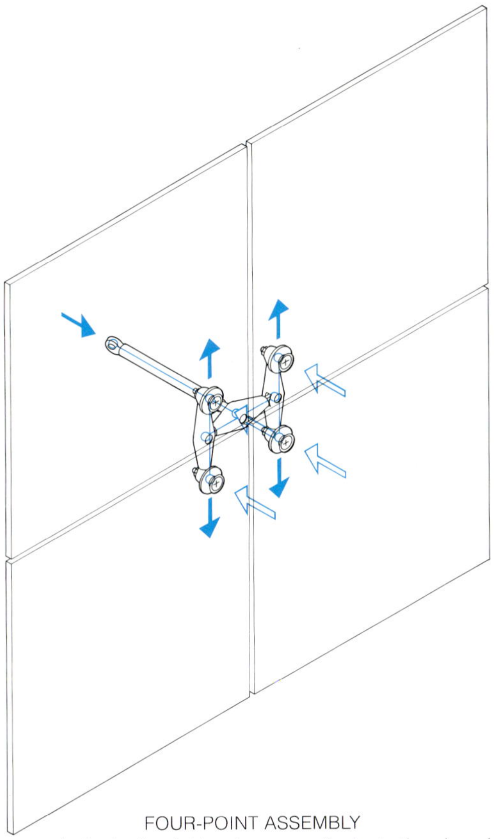

FOUR-POINT ASSEMBLY
Opposes only the horizontal loads perpendicular to the glazed plane. The weight of the glass is transferred to the upper panes.

The one-, two- or four-point assemblies.

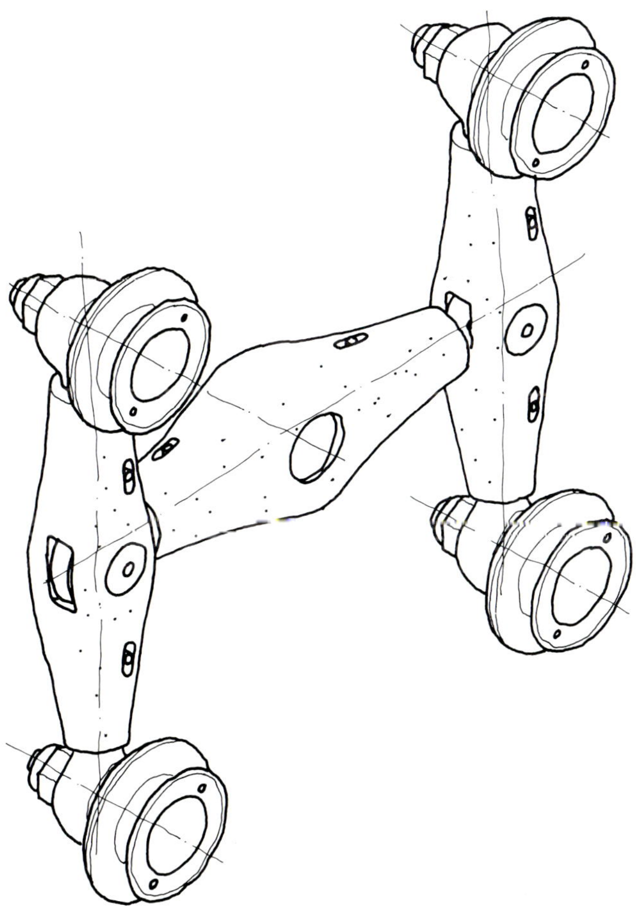

The H-shaped four-point assembly.

lateral loads in the plane of the sheets of glass. The following diagrams show how these assemblies work. Their location in a panel is indicated in the general diagram on page 40.

Details of the four-hole connection

The four-hole connection takes the shape of an articulated H. The accompanying series of drawings describe this component in detail. The four eyebolts of this connection receive the standard articulated bolt described in Chapter 1. They are screwed into a vertical female threaded casting. It is this assembly that takes the weight of the glass sheets. The screw action enables the adjustment of the assembly on site to take into account the manufacturing and assembling tolerances. The distance between the holes drilled in the glass sheets is increased by unscrewing the eyebolts on the vertical casting.

The eyebolt stem is threaded for part of its length and smooth for the rest. Each of these two parts has a specific function: the threaded part takes the vertical load of the weight of the glass, whereas the smooth part works in bending (caused by the lever arm action due to the eccentricity of the plane of glass in relation to the eyebolt centre-line axis). The threaded part is bad at resisting bending: the threads cause stress concentrations in the metal which favour crack propagation. On the other hand, the smooth part of the stem is located at the mouth of the casting where the bending loads are greatest, and so prevents such a stress concentration problem from arising.

Each eyebolt is locked into the female casting by a spring pin; this is made possible by chamfering one side of the stem-threaded section to provide a flat edge. An oval hole is provided in the corresponding female cast component. When the eyebolt is screwed in to the correct length, the pin can be inserted in this oval hole; it rests against the flat part of the stem, thus locking the assembly. Without this locking action, the eyebolts could rotate freely, causing the glass sheets to shift out of plane.

The vertical casting fits directly onto an eyebolt which is screwed into a casting identical to the first one but positioned horizontally. This connection must articulate when working in bending; that is to say, it must not stick when subjected to bending loads and consequently prevent the H-shaped connection from moving freely. The detail designed for this purpose consists of a pin fitted into the hole of the horizontal eyebolt. For assembly purposes, it is made in two parts firmly tightened onto the eyebolt by a connecting screw.

Sketches of the four-point assembly.

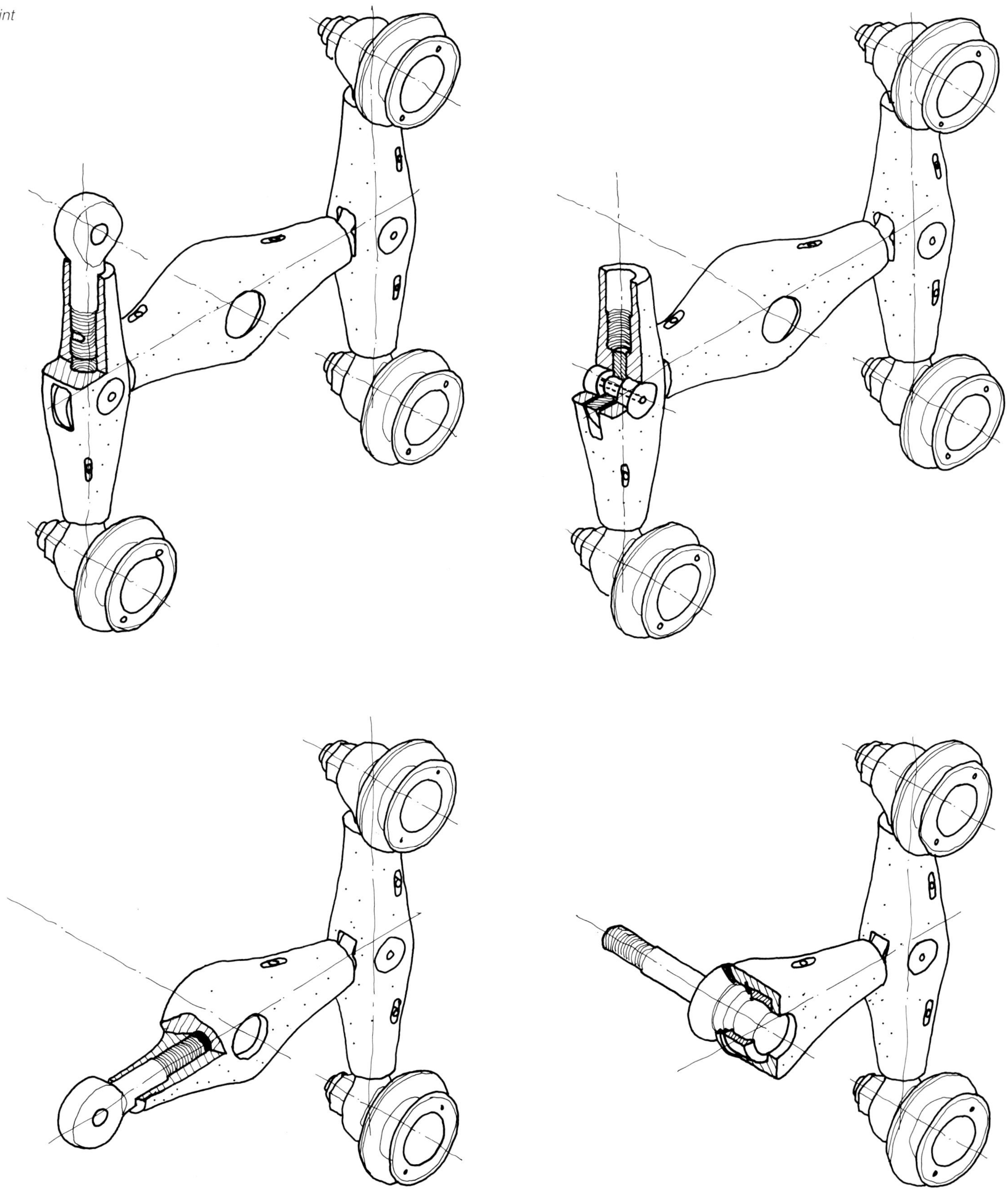

The pin and the corresponding hole in the vertical casting are both machined to close tolerance. Mechanically this assembly predicts that the twisting loads from the vertical casting will be transferred to the horizontal eyebolt by the pin rather than by the flat faces of the eyebolt head. To guarantee the freedom of rotation of the assembly, thin PTFE (polytetrafluoroethylene) washers, a low friction material, are fitted between the flat faces of the eyebolt and the vertical casting.

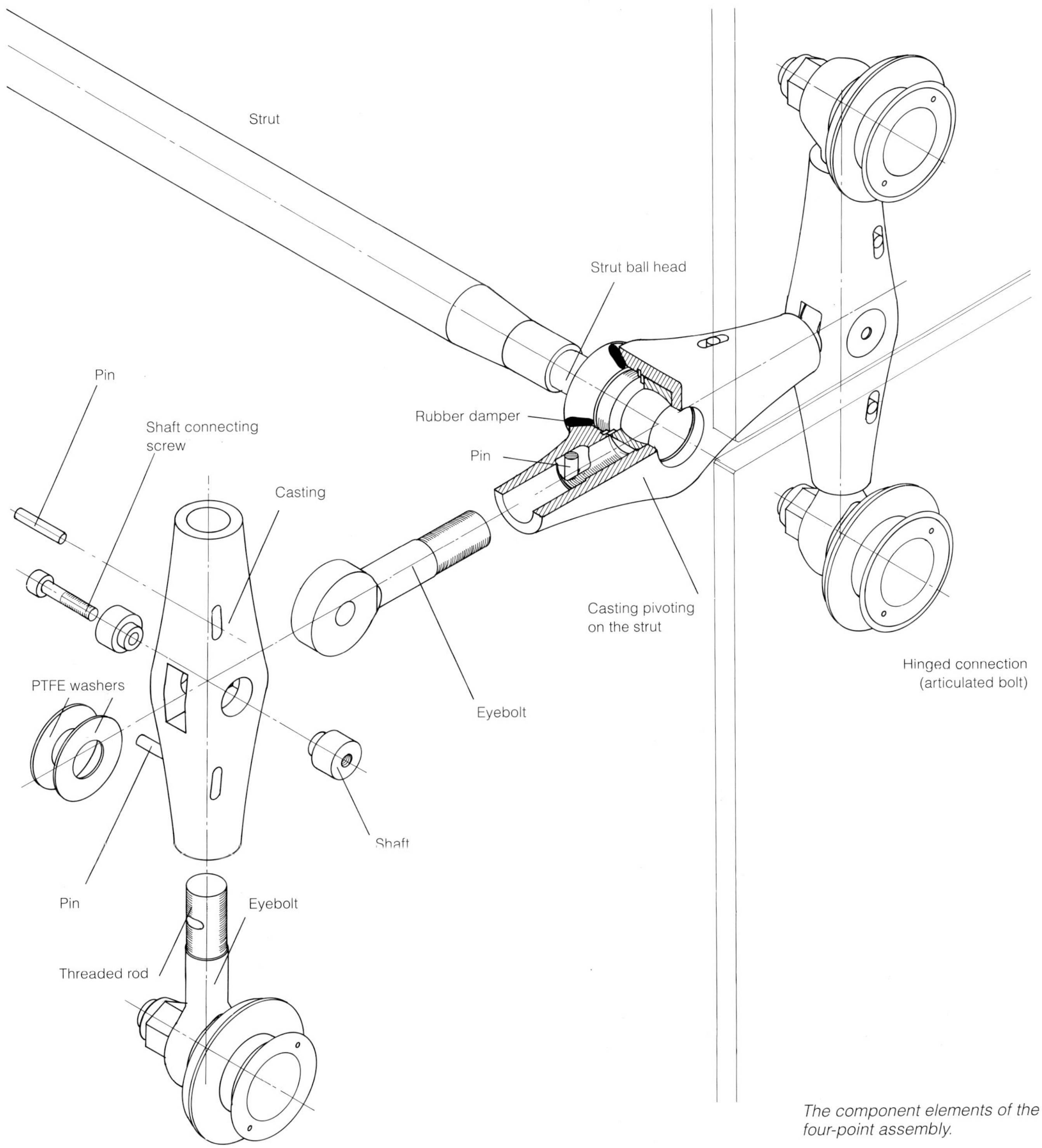

The component elements of the four-point assembly.

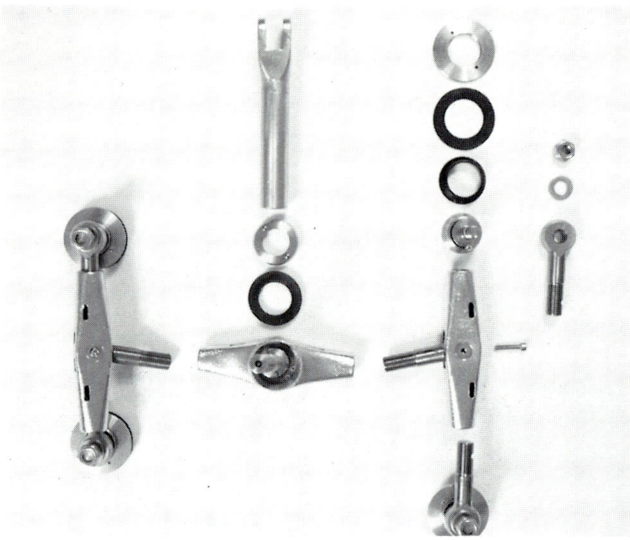

The four-point assembly:
- *front view showing component elements;*
- *view showing complete assembly.*

The horizontal eyebolt is then fitted in the horizontal casting in exactly the same manner as those in the vertical casting. It is also locked by a pin. This 'equivalence' between the horizontal and vertical castings is important when the four-hole connection is fitted to the two-hole connections located on the perimeter of a panel; in such cases, the eyebolt that receives the articulated bolt fits directly into a casting hinged onto a small strut connecting the whole assembly to the structural frame. The casting is oriented horizontally if the connection is at the top or bottom of the panel, and vertically if it is at the sides.

The horizontal casting of a four-hole connection is attached to the end of the cable truss struts by a spherical bearing connection. This connection allows the designer to predict that no movement of the strut can ever apply a twisting action at the glass support points. For instance, if someone hangs from a cable truss, the strut bends downwards; the connection ball rotates in its housing and no moment is transferred to the other elements of the four-hole connection.

The standard spherical bearing has a threaded stem and fits into a precision hole machined in the casting. A standard circlip retains the housing of the spherical bearing within the hole.

Finally, a rubber ring or shock absorber is positioned between a shoulder on the threaded stem and the casting. This ring acts as a self-centring spring for the stem. Given the high number of articulations which are necessary for mechanical purposes, the assembly must not be too flexible otherwise it will move under the slightest asymmetric or unforeseen force. The rubber ring ensures that the assembly endeavours to find its theoretical geometry. The articulated bolts are fitted with the same type of flexible O-ring. Finally, it plays the additional role of shock absorber in the event of breakage of the glass sheet. These rings prevent sudden shock loading from being transmitted through the system.

Watertightness of the glass panels

The seals between the glass panels are made of silicone. This material is used either as a mastic or as extrusions glued together with mastic. The mastic is used on the small joints between glass sheets in a glass panel, and the extrusions are used on the larger joints between panels. Silicone adheres perfectly to glass when applied according to the following precautions: the glass must be absolutely clean and dry; the area of the joint must not be too wide – this could prevent its total polymerization. Under these condi-

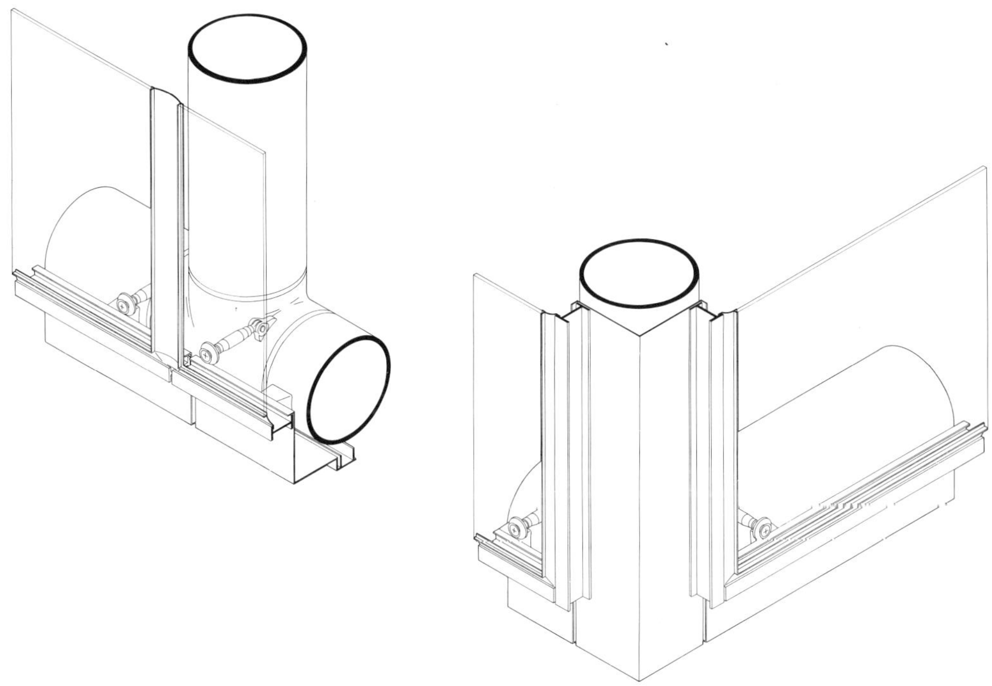

The watertight seal between the panels is achieved with extruded silicone joints.

tions, silicone has a remarkable adhesive power, a fact which was confirmed by the tests carried out for this project: the joints expand up to four times their original size, and in case of failure they are cohesive and not adhesive, which means that the failure will occur in the joint itself and not on the glued surface. The extrusions between panels have been specially designed for the Serres at La Villette. The joint must satisfy three requirements: firstly, it must allow the panels to move in relation to one another; secondly, it must not protrude from the glass external skin; and, thirdly, it must be watertight. The extrusion process involves forcing the basic material through a die. The extruded joint is translucent, which minimizes its presence: it must neither suggest a frame for the suspended glass system, nor disturb the perception that the glass hangs directly from the structural steel frame. Since this type of application for extruded joints was unknown before the Serres project, the joint has been rigorously tested in an accelerated aging device which subjected it, under load, to alternate cycles of humidity, frost and heat.

Safety in accidental breakage cases

The possibility of a fracture of the glass must be taken into account. The breakage of a glass sheet would cause immediate changes in the load path of the glass suspension system. Should the upper sheet in a row of four sheets break, the remaining sheets will no longer be supported; they will transfer their loads to the adjacent sheets of glass through the horizontal casting of the four-hole connection and by producing shear stresses in the silicone joints. The loads applied at the suspension points of the three other rows will also be increased. The sketches given on page 51 show a few possible load permutations following the fracture of one or two sheets of glass and the ensuing change in the load path. Furthermore, a sudden change in the load path may cause a violent shock at the suspension points.

A sudden load is far more likely to cause a fracture than a gradually applied one. The shock absorbers and springs previously described reduce the effects of such shocks. The tests carried out on the prototype included an analysis of the system's strength in the case of breakage of two of the four top sheets of glass.

Process: model and prototype

An innovative design must give confidence to the designers' team and to all those involved in the project – the client, the contractors, the checking authority – as well as to the general public. The purpose of

From the inside, where four sheets meet, a cable truss, a four-point assembly and a myriad of shadows projected onto the panel.

the model prepared during the preliminary project phase was indeed to show the public that a suspended glass system with cable trusses was a tangible and achievable concept. This model was presented at the exhibition 'Architecture et Industrie' at the Pompidou Centre in 1983. The details were not final at the time, but the principles of suspended, point fixed glass were clearly expressed by the model; it was built to scale and was able to convince the spectator of the project's feasibility.

After completion of the drawings and before the implementation phase, another full-scale prototype was built and its resistance to wind, rain and accidental breakage were laboratory tested.

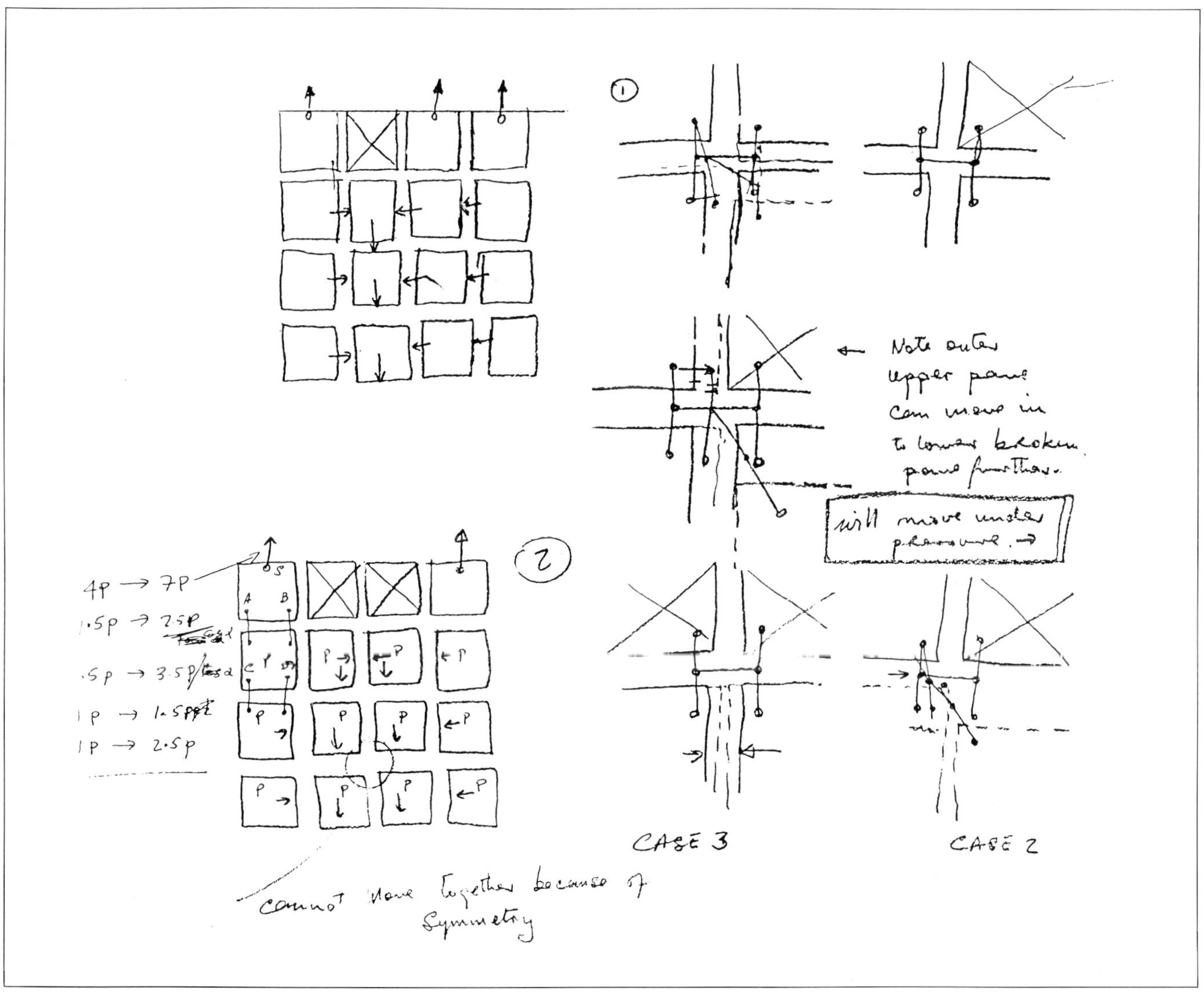

Above:
On show at the Pompidou Centre in 1983, a model of the cable trusses.
Left:
Diagrams of the permutations of loads on the inside of a panel during failure of a glass sheet.
Below:
Trials on the prototype; here the two sheets at the top left are broken.

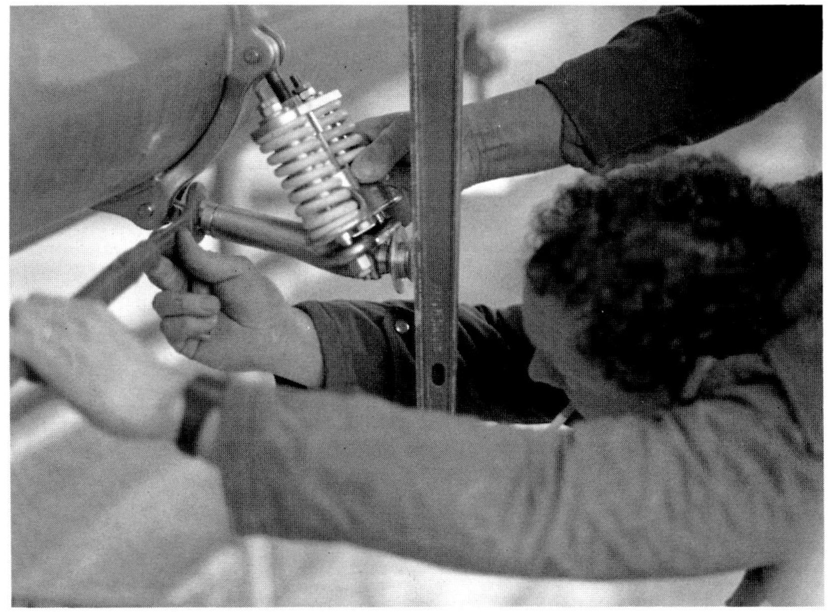

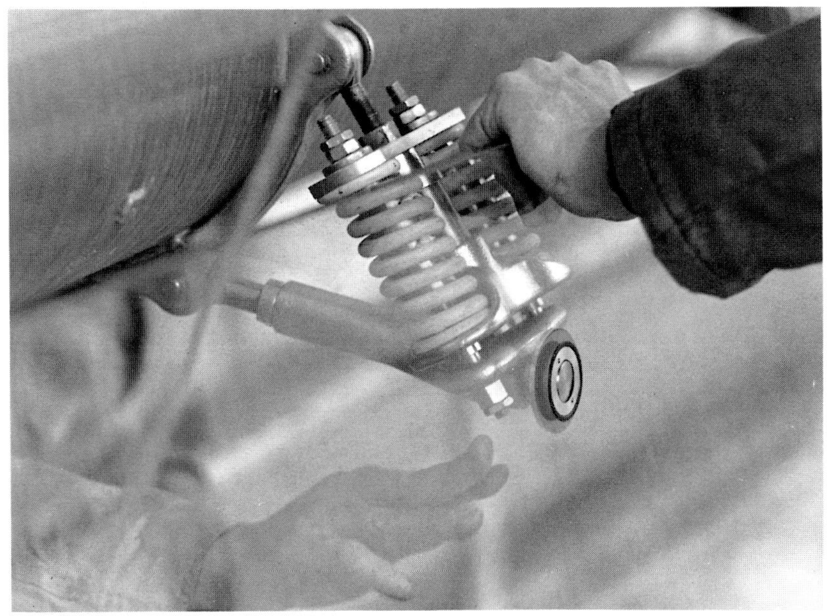

Left-hand page, from left to right, and top to bottom: Mounting the suspended glazing to the framework from the scaffolding. The sprung suspension system is positioned then fixed to the framework. The glazed module is then offered up and suspended from this assembly. Lastly, the other panes in a column are fixed, one after the other.

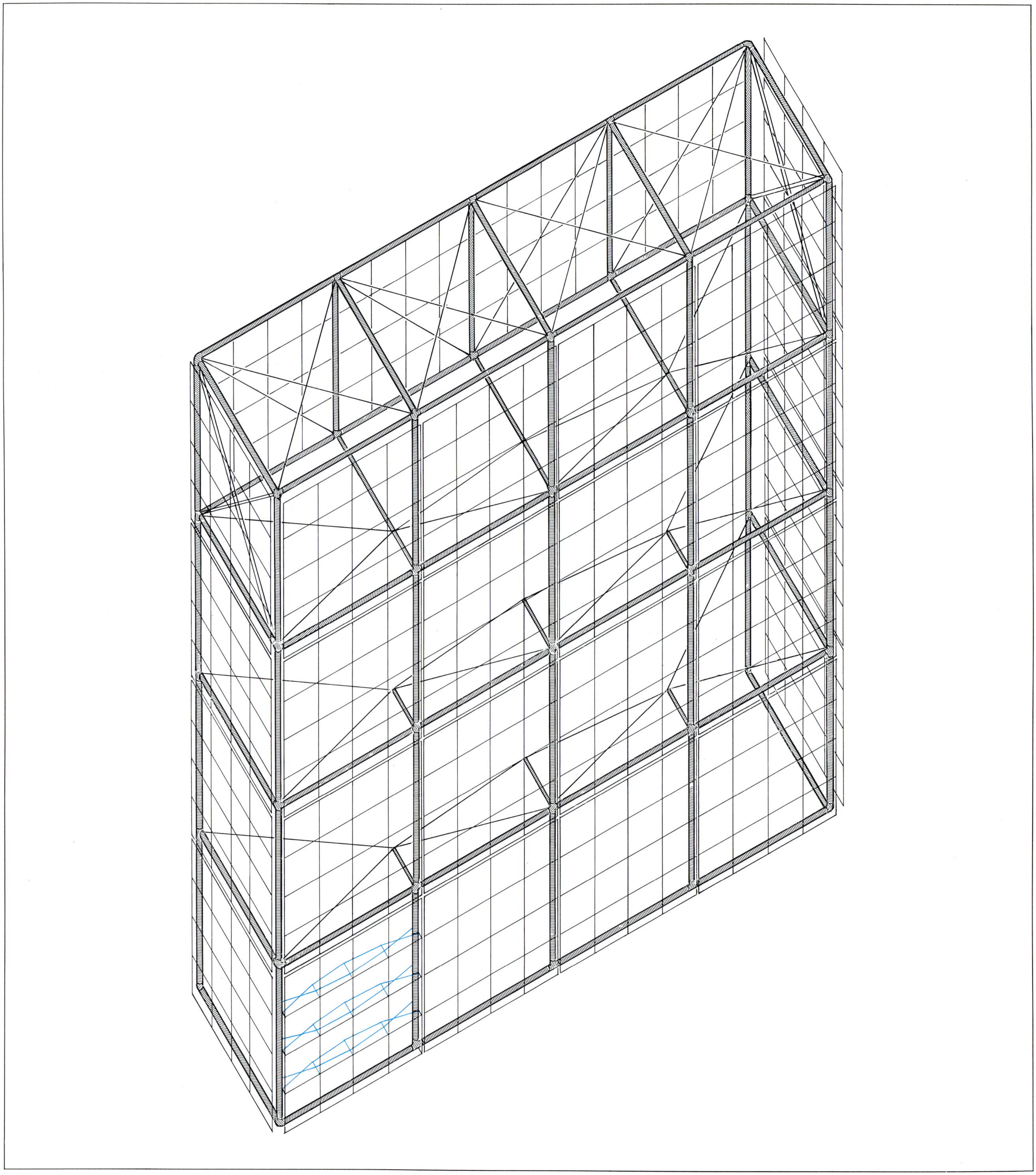

CHAPTER 3

THE CABLE TRUSSES

Refining the solution

At the time of the preliminary scheme, inspired by the Willis Faber & Dumas building, the design of the glass support system was somewhat problematic. The glass fin and patch plate clamping attachment system initially intended is very transparent when viewed straight on, but acquires a density and loses its transparency when viewed from an angle. Despite the fact that all of the façade, including the stiffening mullions, are in glass and are, by definition, transparent, the visual presence of the connections – and the fins with their reflections – destroyed the impression of real lightness.

A more subtle solution had to be found, whose physical presence would minimize all obstruction in the visual field. There was also a fear that, seen from the outside, the large assembly plates located at the

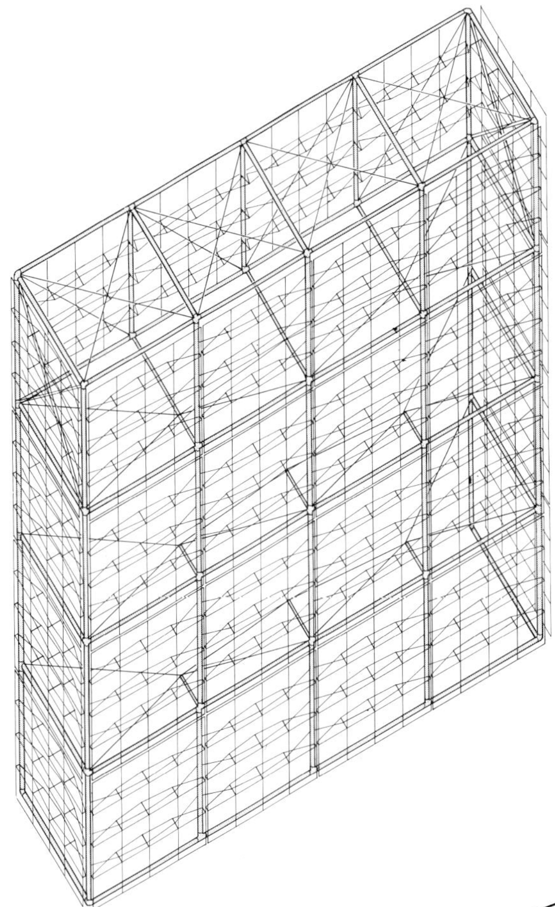

The stability of the glazed plane is ensured by the cable trusses, arranged horizontally and fixed to the framework.

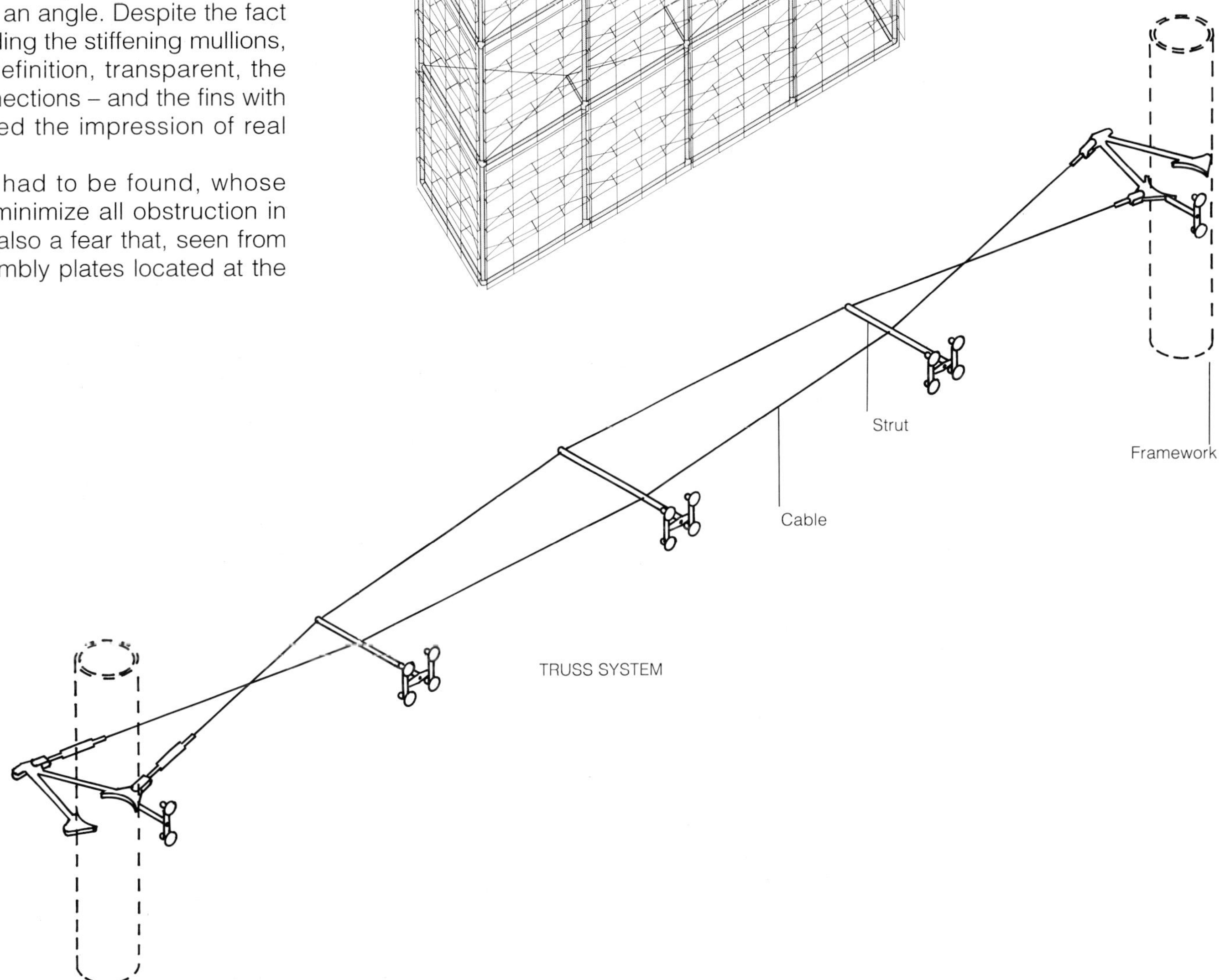

3 THE CABLE TRUSSES

The cable trusses do not diminish the transparency of the glazed plane.

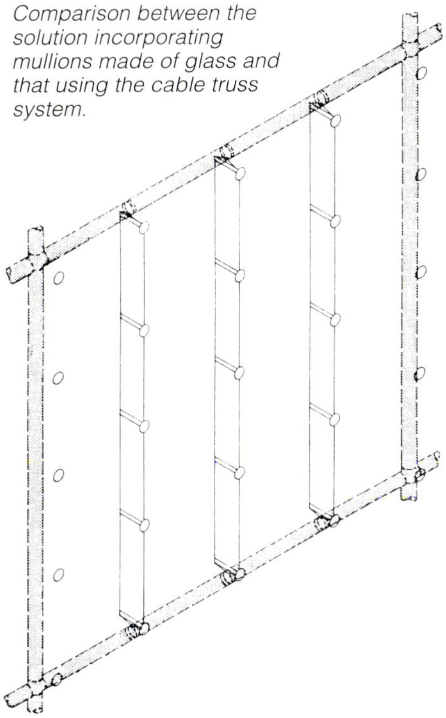

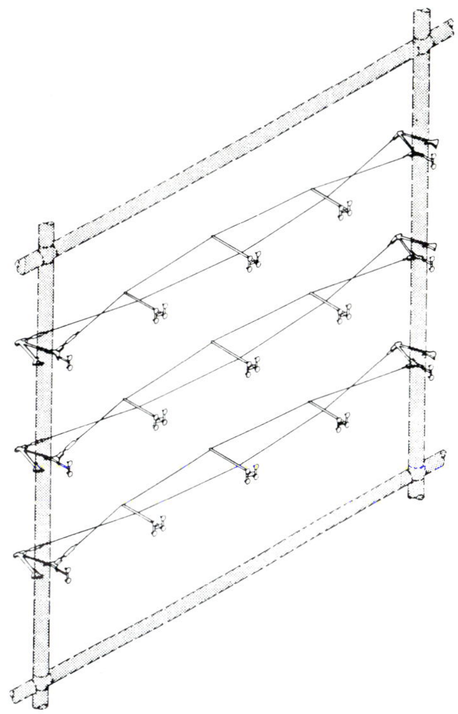

Comparison between the solution incorporating mullions made of glass and that using the cable truss system.

The truss cables for the Serres were selected from a yacht fittings catalogue.

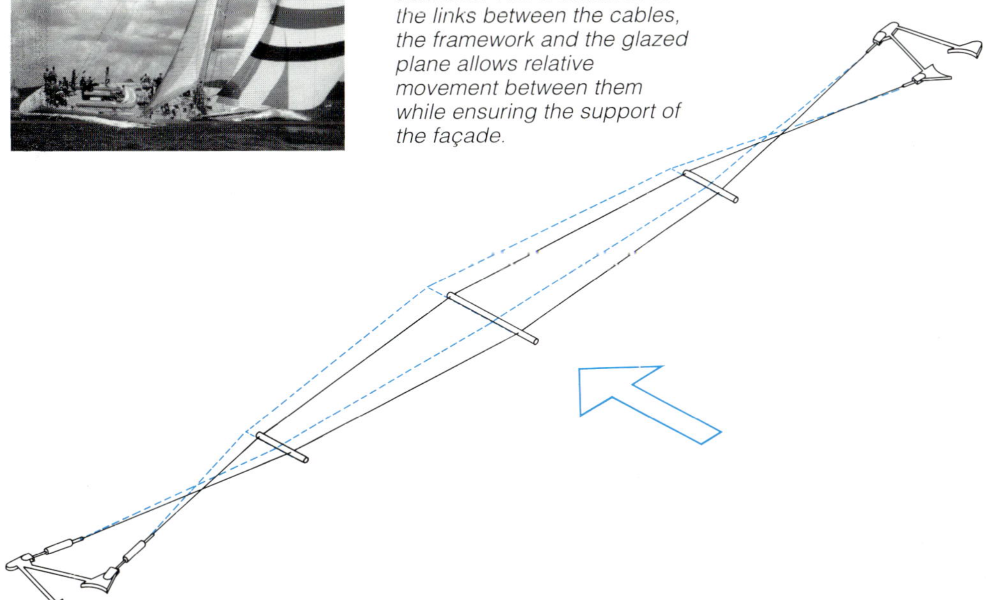

Below:
Under the influence of the wind, the outer truss cable slackens. The articulation of the links between the cables, the framework and the glazed plane allows relative movement between them while ensuring the support of the façade.

corners – the clamping patch plates – might reduce the impression of a suspended assembly. All these considerations led us to replace the glass mullions with a cable truss system anchored at the back of the structural frame. We soon realized that the cable trusses could be horizontal, which would accentuate the feeling of lightness of the structure and would not obstruct a panoramic view of the park.

Description of the truss system

The cable truss is the only horizontal support system for the glass and it is identical throughout the entire façade. Its structural performance is discussed later in this chapter. The cable truss consists of two single strand cables, 12.7 mm in diameter and made of 19 wires (selected from a yacht fittings catalogue) complete with turnbuckles and fork ends for tensioning and fixing. Their connection to the tubular structure and to the glass attachments has a certain amount of freedom to enable the glass and the truss to move independently while maintaining the lateral support of the façade against the wind.

The truss has two cables, parabolic in shape, which are tensioned one against the other. The shape is achieved by the use of struts which keep the cables apart. The ends of the cables are supported by V-brackets attached to the main frame columns. The total depth of each truss is 600 mm. The distance between the two cables is therefore 600 mm at the anchorage points of the cable ends and at the centre; they cross over at a distance of 1.20 m from each cable end.

The cable trusses are positioned horizontally inside the main frame. This means that, taking into account the offset of the main frame in relation to the plane of glass, they are 300 mm from the glass plane at their closest point. Struts, which line up with the struts in the cable truss, hold the glass in a given position relative to the cable truss. These struts are fitted with horizontal hinges located just behind the plane of the glass and at their connection with the cable truss. This allows a horizontal independence between the glass and the truss in directions parallel to the glass. This is necessary since the truss will change its horizontal shape and position under load.

The cable trusses are prestressed to 2 tonnes per cable. Under maximum load, one of the two cables can 'lose' its pretension, the load being taken by the other cable alone.

The reasons why this level of prestress was selected and the way in which the prestressed truss system

Lateral deformations in the truss do not affect the glazed plane.

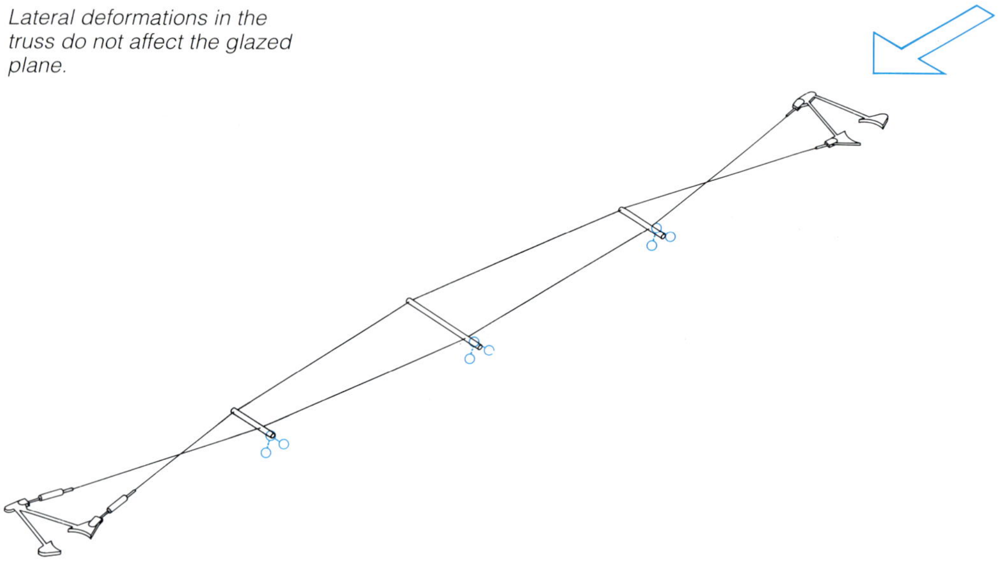

works will be explained later. Before that it is important to understand the nature of the loads that must be taken by the cable truss system and the glass.

Wind loading

The variable nature of wind loads

The horizontal truss system and the main frame primarily resist wind loads. These loads vary according to the direction of the wind and can become high inward pressure or outward suction on the main glass wall. Pressure and suction can also be applied to the side walls; they also produce forces parallel to the main building which may be coupled with the inward pressure or outward suction already mentioned.

The problem of wind forces must be considered from a statistical point of view: light and frequent wind

The horizontal layout of the trusses minimizes their visual impact.

The different wind loadings on a cable truss:
- *positive pressure;*
- *negative pressure;*
- *combined effects of positive and negative pressure.*

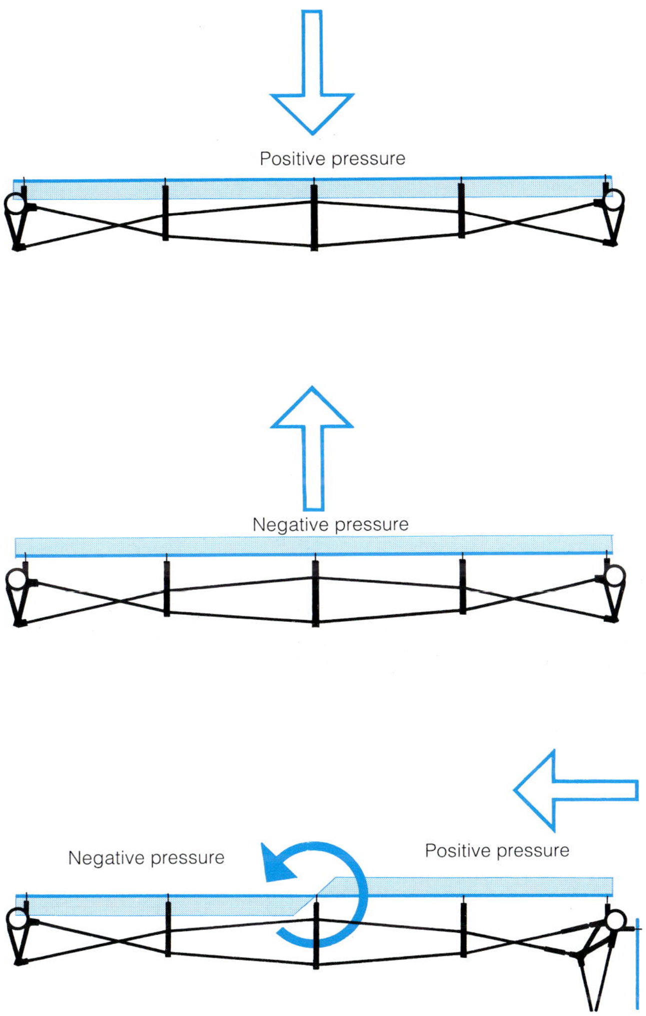

forces can be predicted on a daily basis, slightly larger forces on a weekly basis and other, more serious forces, on a monthly basis. The actual wind forces for which the building at La Villette has been designed are those likely to occur at most once in a hundred years. The design satisfies, in fact, two different and seemingly contradictory hypotheses: it permits deflection, but at the same time limits it. It is a question of degree – how much deformation can be allowed? Normally, a threshold of permissible deformation is established for frequent wind loads that is compatible with the glass or aluminium cladding systems involved. The joints in these systems permit degrees of movement that are compatible with the threshold. Above the threshold the cladding can be damaged.

At La Villette, however, quite a large deformation threshold can be permitted because the glass fixing system allows a considerable degree of movement because all of the connections are articulated, notably the glass bolts.

However, there is a further limit to deflection, even if the cladding fixing system permits a considerable amount, which is the psychological limit, that is the point at which the movement disturbs the general public visiting the museum.

The design of the cable trusses uses the stiffening device of prestressing, which limits the deformation for statistically frequent loads. If, however, the loads during extreme conditions are above the psychological threshold, the prestress can be overcome and the deformations become quite visible, even though there is no risk of failure.

According to the French regulations governing structural designs, all structures must be designed to withstand normal and extreme wind loads. This requirement has led to a distinction between frequent forces and those occurring 'once in a hundred years'. It is in this context that the system designed for the Serres at La Villette is rigid or of low movement under normal loads, while able to move greatly under extreme loads.

Patterns of wind load

The wind can blow on the façade from any direction, and each direction implies a different pattern of loads. Obstacles in the path of the wind before it reaches the building can cause it to gust and change direction suddenly. For this reason any pattern of wind load is possible on the façade. The pressure from the wind and the maximum pressure which a glass structure must be able to withstand are defined by the regulations and on the basis of the research undertaken on the interaction between the shape of a building and the wind direction.

For La Villette, standard patterns of wind loads most likely to cause the worst load conditions on the glass and on the truss are used.

Some of these are shown in the diagrams on this page, which analyse the pressure on any kind of wind truss system. In addition to external pressure, the system may also be subjected to internal pressure. The latter is always uniform and is usually approximately one-third of the maximum outside pressure. It acts simultaneously with the external pressure and results from the cladding not being airtight or from open doors or windows, which cause the building to pressurize. This pressure can be positive or negative (depending on the location of the openings); internal and external pressures must be combined to study the worst possible consequences for the Serres.

Effects of temperature

There is another environmental factor to be considered. It is the effect of changes in temperature. The various elements of the Serres building change temperature at different rates according to their mass, their surface area, whether or not they are in the sun, etc. Any change of temperature occurring in the various elements of a prestressed structure can imply a prestress loss or gain. This factor must also be examined.

How a cable truss resists loads

Generally speaking, external loads applied on a beam are translated into internal forces which act on each of its sections. These forces are of two types: moment and shear. The moment about a given axis is the force that prevents rotation of a section around it. Shear in a given direction is the force that prevents a section from sliding in this direction.

Typically in a horizontal beam under a vertical downwards load (Figure 1) the moment, called the 'bending' moment, is taken by the upper fibres going into compression and the lower fibres into tension (Figure 2a). The shear, equal to the vertical reaction applied at the support on that side of the beam, is taken by the web of the beam (Figure 2b). It is the combination of these two forces that enables the loads to be transferred to the supports.

It is the same in a three-dimensional beam consisting of horizontal, vertical and diagonal members, called a 'truss girder' (Figure 3a), where the two horizontal chord members, respectively in compression and in tension, take the bending moment, while the diagonals take the shear (Figure 3b). The forces are therefore transferred by axial loads in the members.

In the more particular case of the cable truss (which does not have diagonals), the shear is taken by the curvature of the cables according to the same principle that applies to a chain under vertical loads (Figure 4a). A chain fixed at both ends deflects under its own weight: equilibrium is achieved in a configuration called a catenary curve. In a chain, the moment is the result of the cable tension and of the horizontal force applied to the cable ends. This horizontal force is equivalent to the compression which would be applied to a normal beam. The shear is taken by the curvature of the chain, which provides a vertical component to

BEAM

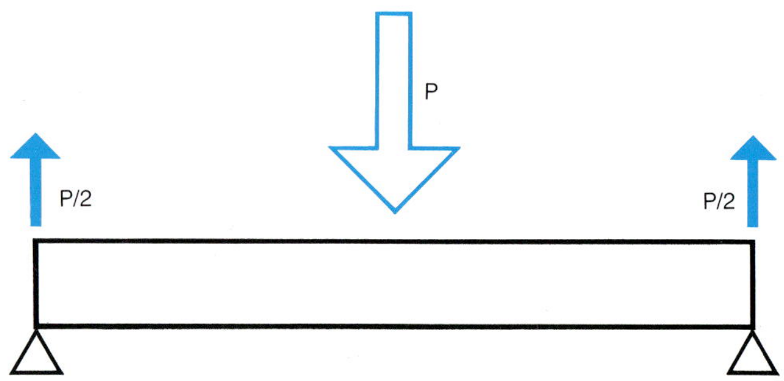

Figure 1
A vertical load P is applied.

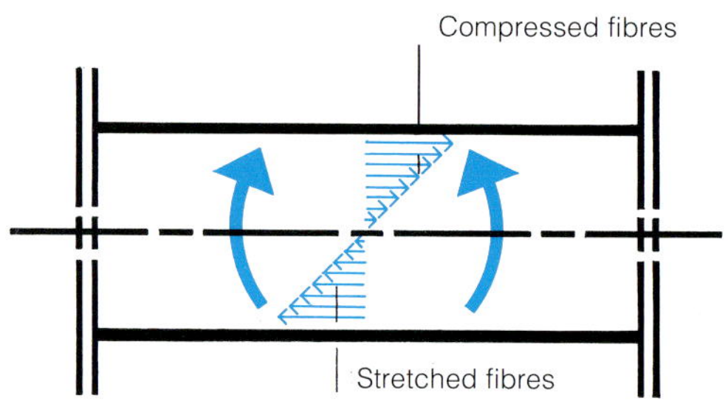

Figure 2a
A bending moment is created.

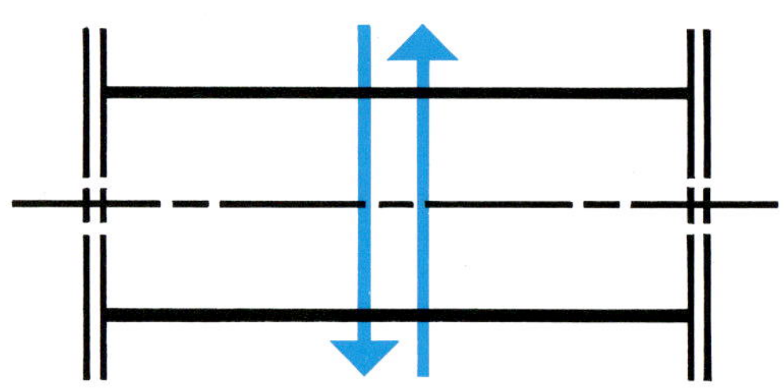

Figure 2b
An opposing shear is created in the core of the beam.

TRUSS

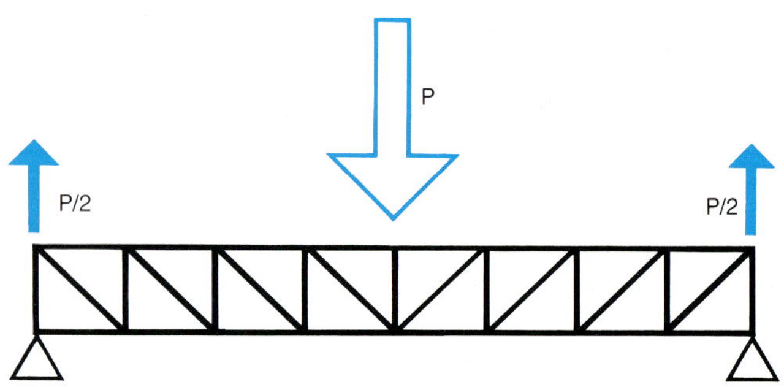

Figure 3a
A vertical load P is applied.

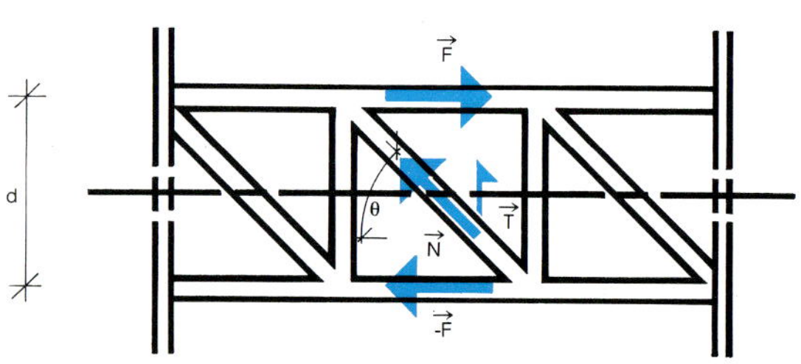

Figure 3b
The bending moment is taken by the two outer members:
$M = Fd$.
The opposing shear is taken by the diagonal members:
$T = N \sin \theta$.

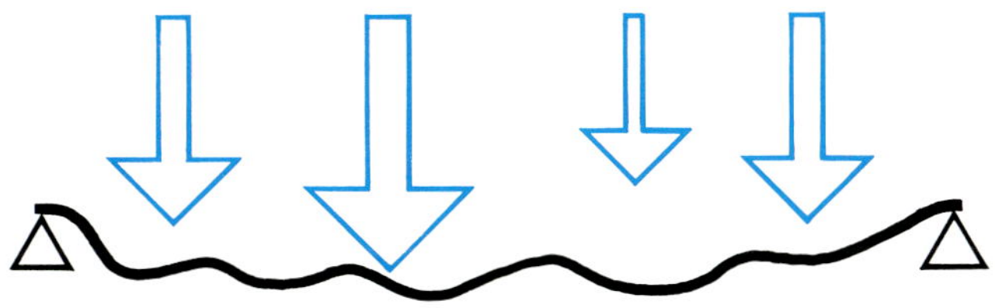

CHAIN

Figure 4a
Varying vertical loads are applied.

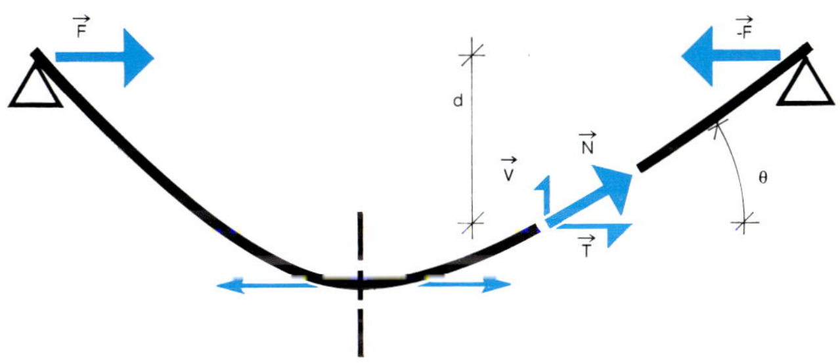

Figure 4b
The moment is the result of the tension in the cable and of the horizontal force applied at the ends of it: M = Fd. The opposing shear is taken by the curvature in the cable: V = N sin θ.

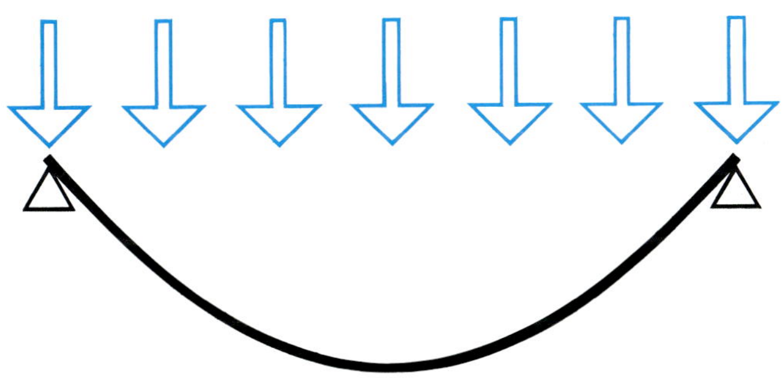

Figure 4c
Under an evenly distributed load, the cable adopts the form of a parabola.

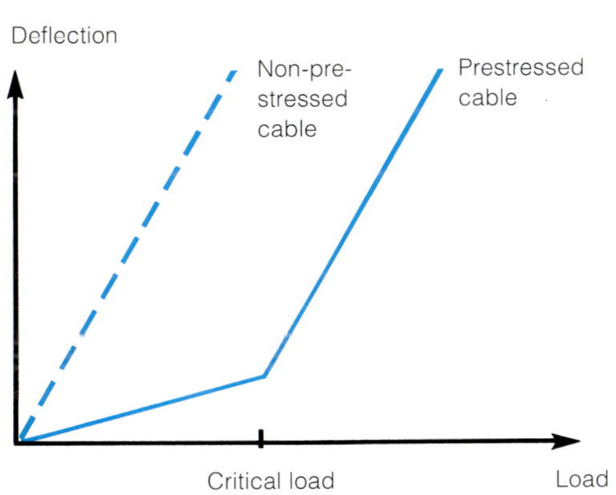

Figure 5
Influence of prestressing on the resistance of cables.

the cable axial tension (Figure 4b). The curvature obtained is determined by the nature and size of the loads applied. The peak of the curve (horizontal tangent) corresponds to the point at which there is no shear. As the catenary curve is in equilibrium with the cable's own weight, applying any new load causes a change in the curvature. In the case of the cable truss at La Villette, the loading taken into account corresponds to wind forces which are evenly distributed in the cable along its entire length (Figure 4c). Subjected to this type of load, a cable takes the shape of a parabola, so this shape was chosen for the trusses.

As has been seen, moment is taken in the same manner as in an ordinary beam: cable in tension and perimeter structure in compression. In the cable truss system, the two cables act jointly as two catenary curves. Under a given load, the tension increases in one cable and decreases in the other. It is therefore necessary to prestress the cables so that they remain always in tension. The shear force is taken by the shape of the cable truss. If the loads are evenly distributed, there is no change of shape because the parabolic curve configuration of the cables already corresponds to an even loading.

On the other hand, in all other cases (asymmetric loads) the truss changes shape to accommodate this load change. This geometrical change is minimized due to the fact that the cable trusses are prestressed. The geometrical deformation becomes relatively important when all the prestress has been overcome in either of the two cables (Figure 5).

This need for geometrical change in the truss is the main reason why hinges have been included in the connections between the trusses and the glass; thus the deformation of the truss does not endanger the glass. This ensures predictable behaviour of the cable truss under loads and also of the glass system.

Advantages of prestressing

There is a simple and obvious advantage. Instead of one cable resisting the load, there are two cables in tension relative to each other. As one cable elongates with an increase in tension, the other shortens as it loses tension. Obviously the efficiency of the system is doubled by this effect. Another advantage comes from the stiffening effect of the prestress itself. This is easily understood when considering a wire pulled taut horizontally while a vertical downwards load is applied to its centre (Figure 6a).

It will be seen that an increase in prestress reduces the final deflection (Figure 6b). This is because under

PRESTRESSED WIRE

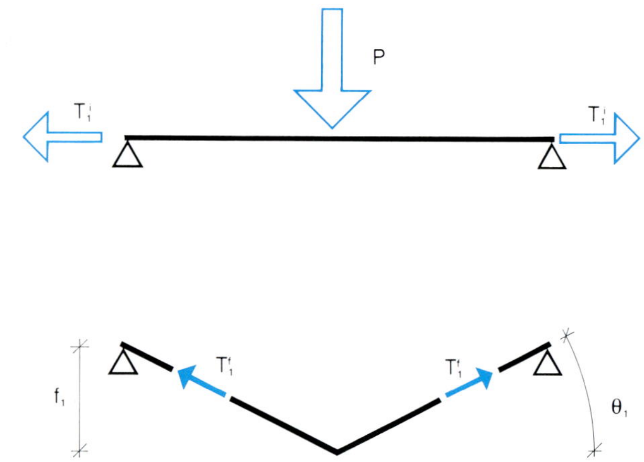

Figures 6a
A vertical load P is applied. The cable then undergoes a deflection f'_1 : $\sin \theta_1 = P/T'_1$.

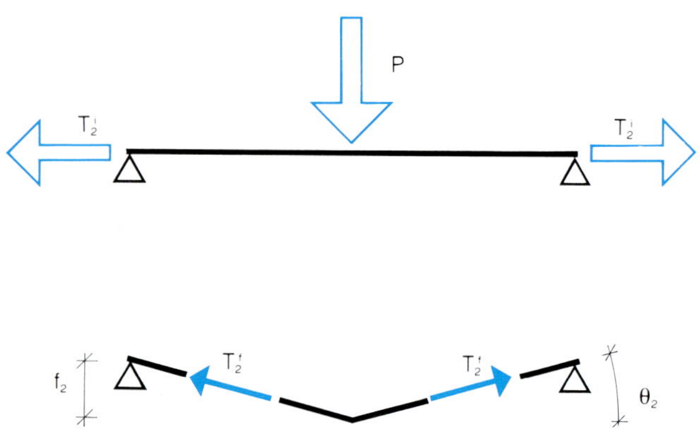

Figures 6b
The same load P is applied. The more highly prestressed cable undergoes a smaller deflection f_2 : $\sin \theta_2 = P/T'_2$.

constant shear, the angle necessary to take the shear will be reduced as the prestress T of the wire increases. The cable truss prestress acts according to the same principle. It only comes into action when a major deformation is necessary to take a load. In a normal structure, where the shear may be taken without significant deformation, the prestress effect is far less important.

In the case of the Serres, it is essential that all the factors likely to reduce the level of prestress be predicted: for instance, a change in temperature or an inadequate initial prestressing. At La Villette, allowance was made for a 15°C temperature difference between the cables and the main structure, which can cause a loss or gain of prestress, and a 20% overstress or understress in the cables, which could be caused by the erection procedure. Since the design allows the unloaded cable to go slack under extreme loads, the system remains just as strong, even if the prestress is less than intended. The only consequence of this understress (or overstress) would be an increase (or decrease) in the structure deflection in the load range where the prestress is effective.

Choice of the calculation method

The main calculation problem was that the current analytical methods in most computer programs assume that the deflection of the structure does not affect the forces in the members. This is true for the majority of structures as they have enough members to carry load without significant geometrical change. Furthermore, normal methods of calculation assume that all members of a structure are continuously under load: if they are not in tension, they are likely to work in compression. Well, neither of these conditions applies to the prestressed cable truss at La Villette. We have already seen that on the one hand, the capacity of the cable truss to resist forces depends on geometrical deformations and, on the other, that the possibility of a loss of tension in one of the two cables poses a fundamental problem.

Therefore the calculation methods used were specially adapted for La Villette. These are based on numerical iteration (many successive calculations in order to approach gradually the equilibrium position of the structure), and with the new generation of computers these are easier to perform than in the past. In this respect the Serres structure is a product of contemporary methods (it would have been possible previously, but with much less precision) whose application will most certainly become more widely used in the near future.

Stability in the vertical plane

As has been explained, the two cables of the truss lie in the horizontal plane. There are no members at all in the vertical plane. The cables are 600 mm apart at their support point, they cross over and are again 600 mm apart in the centre of the panel. Because of this cross-over, the central part of the truss could rotate about the horizontal axis through the two cross-over points.

Furthermore there is no guarantee that the (wind) loads applied to the truss by the glazing will remain in the plane of the truss. Although it is assumed that

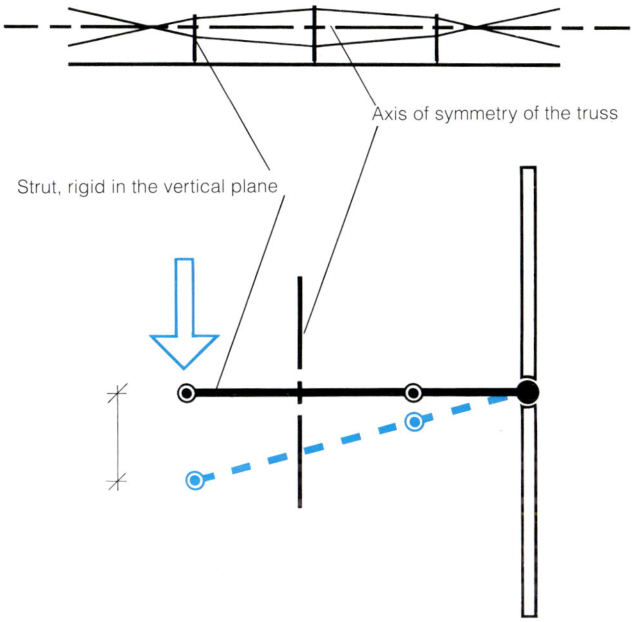

these loads are horizontal,* it is possible that the trusses may not be perfectly horizontal due to errors during the construction phase.

To prevent rotation of the cables about the axis defined by the two cross-over points, the three spacer struts are rigid along their length (for vertical bending) and are only articulated at their connection with the glass (Figure 7). Thus it is the plane of glass itself that prevents any possible rotation. The tension of the cables also helps to maintain the plane of the truss in a horizontal position.

Finally, the glass must be able to move laterally without hindrance from the truss, which requires an additional hinge around the vertical axis of all the cable spacer struts. To confirm the system's performance the calculations were based on the weight of a man hanging from the truss.

Attachment to the main frame

It has already been said that the trusses are independent of one another. This means that the main structure columns had to be designed to support the anchorage of two cable truss systems, taking into account the fact that these trusses are prestressed and have to take horizontal loads. The anchorage system on the columns also had to be designed to withstand a great number of possible conditions – including when two adjacent panel trusses are stressed and loaded or when one truss is stressed while the adjacent one is slack (a condition which could arise if some sheets of glass were broken). The anchorage to the main structure columns is achieved by V-shaped assemblies. These are welded onto the columns before erection on site. There are three different types of assembly (see sketch below): one in the centre of the façade for normal cases (A); one for the corner columns of the façade (B); and a third for the two columns against the main building. The geometry of each component is different, but they are all designed for the same types of load.

*Let's not forget that the weight of the glazing is taken by the glass and by the spring connections.

Elements of the cable truss and manufacturing process

V-Brackets

The V-bracket connection to the tubular main structure presented a particularly tricky geometrical problem. The loads to which the brackets are subjected are quite considerable: 2 tonnes of prestress and up to a total of 4 tonnes under extreme wind load conditions. The problem was how to transfer the loads to the column without the risk of ovalizing it.

In addition, the system's parameters had to be calculated as though the panel adjacent to the one under consideration did not exist, i.e. without relying on a balance between the two trusses. This meant that any load applied to the end of the brackets would apply considerable twisting forces to the column. The effect of the forces had therefore to be reduced to the minimum by applying them tangentially to the column walls; local pinching of the columns due to point forces is thus avoided by the V-shape of the brackets.

To ensure that the whole column, and not just a part

Left:
Figures 7
A cable truss is stable in the vertical plane.
Below:
The three different types of connection of the cable trusses to the structure.

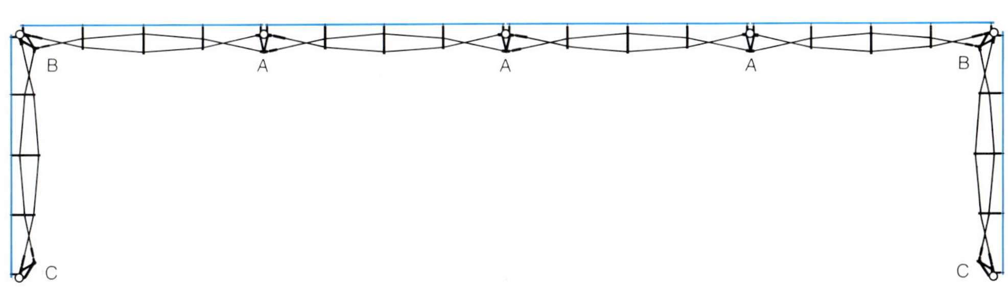

of it, took the load perpendicular to its axis, a reinforcing piece was inserted inside it. The pieces which resolve these complex force geometries required a corresponding complex shape. To meet this requirement, the lost wax casting was employed again because it offered a certain formal freedom, making it possible to offer elegant solutions to these fairly complex constraints.

The struts

Wind loads on the glass are transferred to the truss through attachments located in the corners of the panels. A series of struts perpendicular to the glass transfers these loads into the cable; the struts also act as cable spacers, which allows the system to take its twin parabolic shape. These struts are tubes welded to cast cable clamps. It is critical that these cable clamps remain simple and elegant and are not ugly lumps added to the struts.

The cable clamps

The cable clamps are in two halves: one half ensures the continuity of the strut through the connection, while the other is a capping piece which clamps down onto the first using two standard M8 bolts.

The special components are made using the very accurate process of lost wax casting, as on the suspension system. This process was inspired by an interest in the aviation industry. The design team became particularly interested in the potential for complicated shapes used in some armament mechanisms, fuel pumps and landing gear assemblies used in aviation, manufactured by this process because of its precision, apart from the complexity of the components. The manufacturers collaborated with the designers and were extremely helpful in the development phase of the project.

The lost wax casting process begins with a negative aluminium mould, the shape of which is produced

Below:
Left, detail of the V-bracket fixing the cable trusses to the tubular framework.
Right, principles for transferring the loads from the truss cables to the framework.

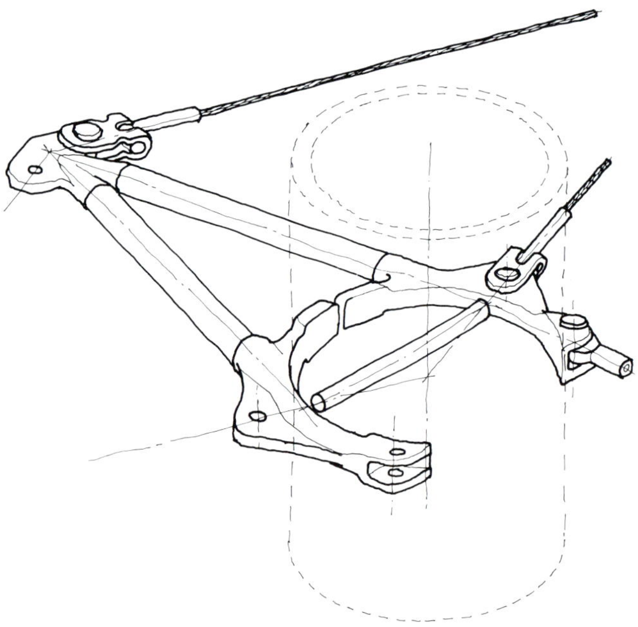

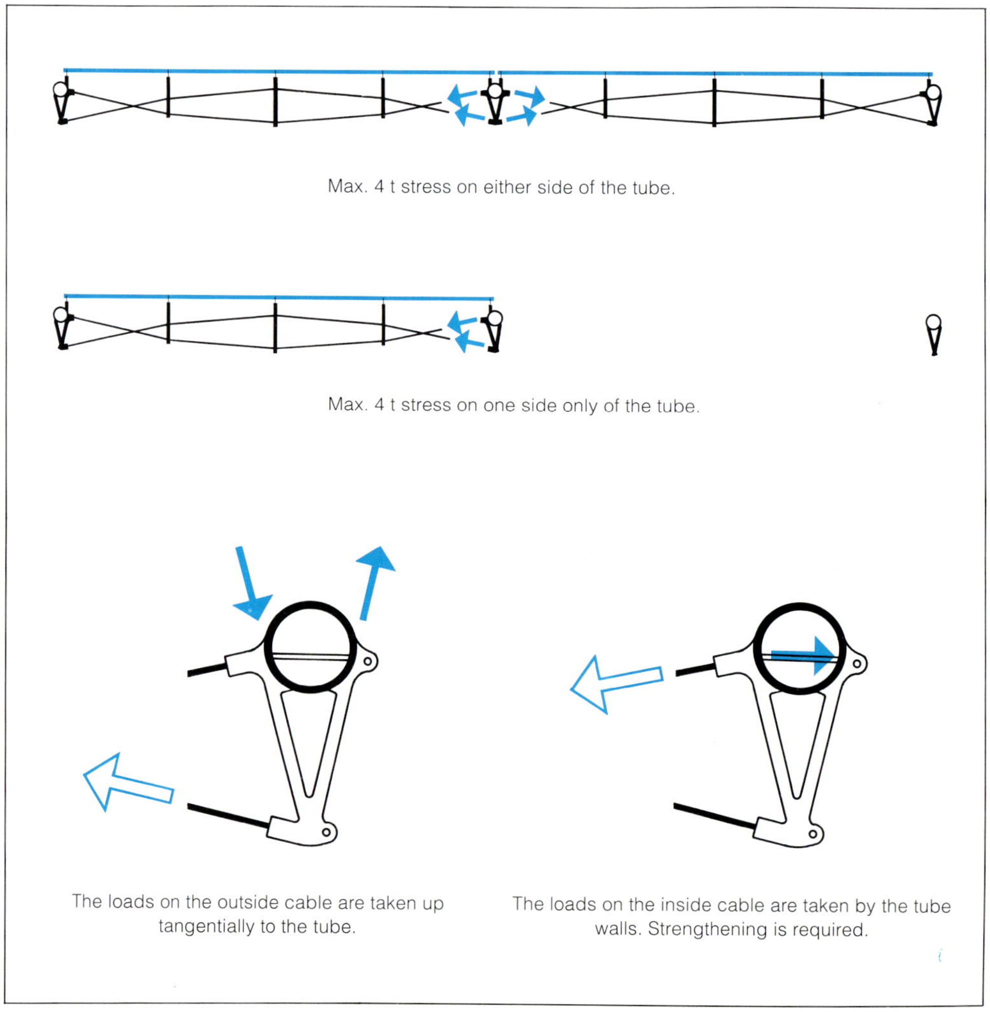

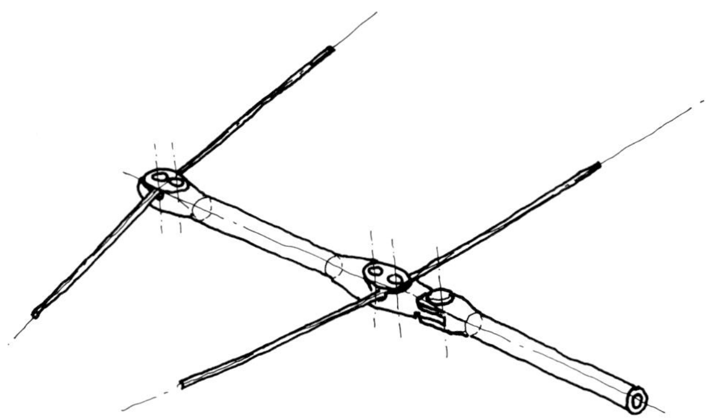

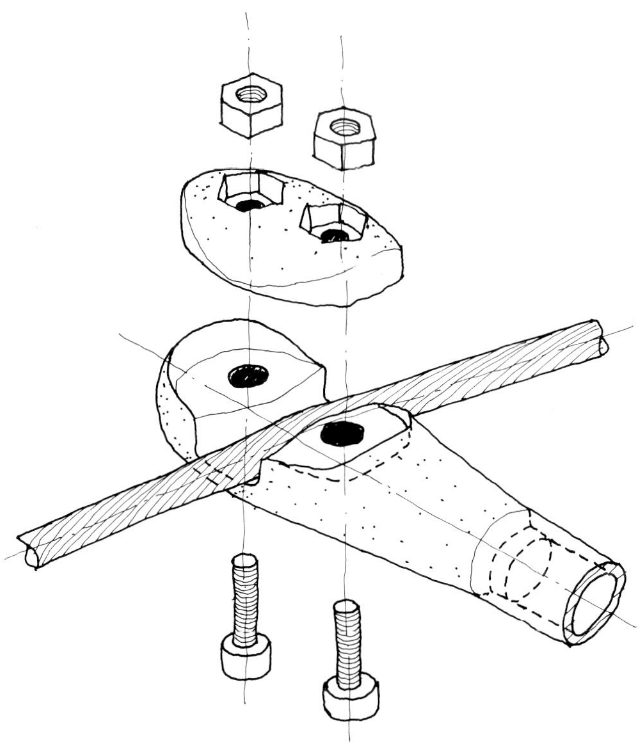

Left:
Above, a small strut.
Below, a cable clamp with its components.

by very precise machining of aluminium blocks. A mould is normally made in two halves. This work must be done by specialist tool-makers skilled in the interpretation of fairly complex three-dimensional shapes from drawings and in their translation into the form of the mould. The designers must therefore prepare very accurate drawings using standard graphic symbols and codes. In some cases, when the required shapes were particularly complicated, the manufacturers built wooden models of the positive shape in order to have a more precise idea of what to achieve. This was the case for the cable clamps of the cable truss.

Once the two halves of the mould are completed, it is closed and liquid wax is injected. This process is repeated for each component required. The positive wax models are then glued to a central stem resembling a bunch of grapes. This bunch is then dipped several times in a ceramic bath to create a ceramic shell around the wax models. The bunch is then baked, melting the wax and hardening the ceramic. The molten metal is poured into the now empty ceramic mould. When cooled, the ceramic is broken off leaving the final casting. This manufacturing process is used when a great many small but precise components are required.

Cables and end fittings

The cables and end fittings are of standard manufacture. They have been selected from specialist catalogues for the sailing and yachting industry. Cables are chosen in preference to rods because they are more flexible. Firstly, a cable taking a given load will have a lower axial stiffness than a rod and will therefore be easier to prestress; secondly, cables are more flexible in bending, allowing continuous cables with cableclamp attachments to be used. All of the cables employed are prestretched to reduce the error in length due to elongation. This facilitates prediction of

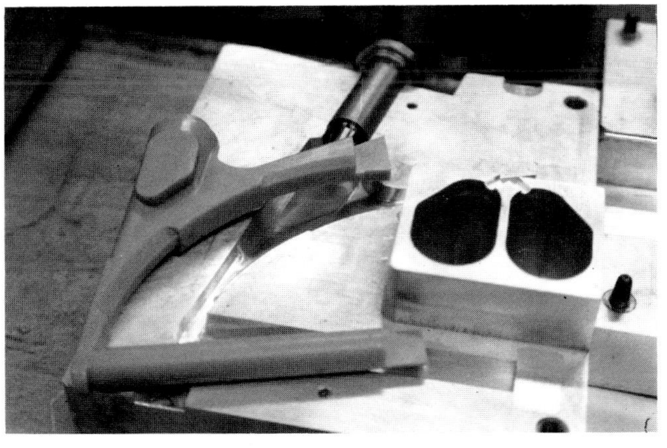

Manufacture of the lost wax castings: on the right the aluminium negative mould; on the left the wax positive.

Top:
Tests on the cable trusses in the prototype.
Bottom:
The cable truss V-brackets were fixed to the tubes of the framework before being brought on site.

their final length after prestress.

As the main stiffness of the truss system comes from the prestressing and not from the cables' resistance to axial loads, the relative flexibility of the cables compared to that of solid rods is not an issue. Additionally, it would have been difficult to curve solid rods to achieve the parabolic shape to which the cables are prestressed. Straight solid rods are, however, used on the main structural frame bracing discussed in the following chapter. The cable end fittings are fixed by the swaging process onto the cable ends. The fabricators use a special rotative hammer machine which plastifies the end fitting metal forcing it into the grooves in the cable strand.

Testing

Cable trusses were made for the prototype of the glass system. During the tests their behaviour was checked under different load conditions right up to the maxima with the most unfavourable level of prestress, to demonstrate that the overall safety of the assembly would not be compromised if the level of prestress was less than the calculated level. The results of these tests were very close to those of our calculations.

Erection

Cables lengthen during the prestressing process. This lengthening can be reduced by prestretching the cables: they are placed in tension until all the wires that make up the cable are bedded together as far as possible. This process does not altogether eliminate the lengthening during the final prestressing stage, but it reduces it to a known value. On site, the cable truss assemblies are pre-assembled on a jig at ground level. There is an assembly bench fitted with all the devices necessary to position the spacer struts used on the cables to a distance of slightly less than 2 m apart; this takes into account the elongation of the cable during the prestressing process and gives a final distance of precisely 2 m. The cables are then assembled onto the struts using the cable clamps described above. The complete assembly of cables and struts is then erected and attached to moulded 'lugs' welded to the awaiting main structure frame.

The prestressing is achieved with a pair of calibrated hydraulic jacking devices which shorten as they stretch the cables. The tension thus achieved is then locked in using turnbuckles.This operation is repeated until the required level of prestress is reached. Once prestressed, the cable trusses remain unstable until the glass is fitted. It is therefore necessary to take certain precautions against their flipping over: safety rings keep the two cables together at the cross-over points, and jigs hold the trusses in a fixed position until the glass is put in place.

Installation of the truss cables on site:

1. *Positioning of the small struts on a jig at ground level (d = the distance between the small struts before prestressing).*
2. *Assembly of the cables which make up a truss system.*
3. *Hoisting.*
4. *Fixing to the framework.*
5. *Prestressing the cables.*
6. *Repeating the operation (three trusses for each panel).*

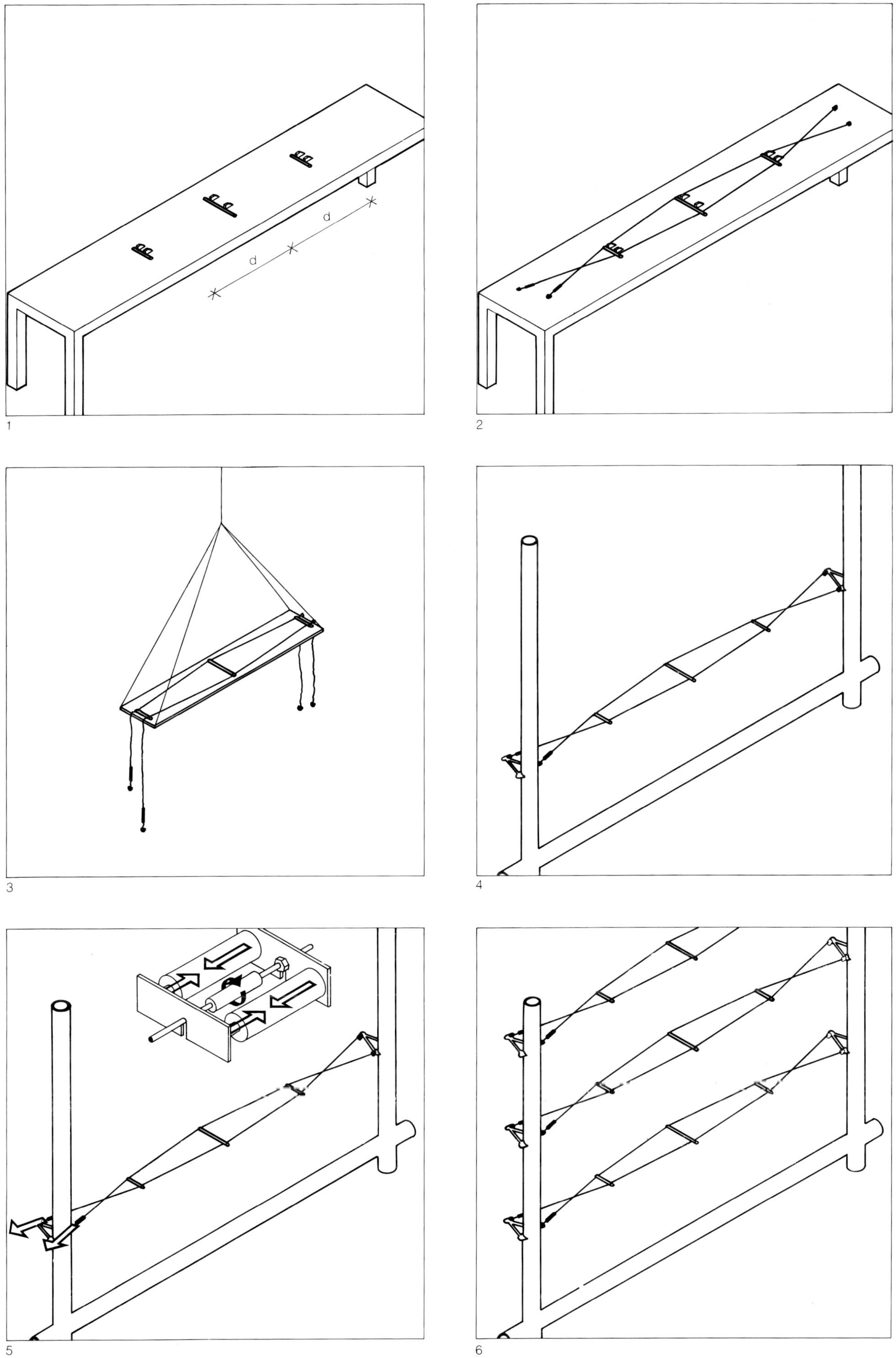

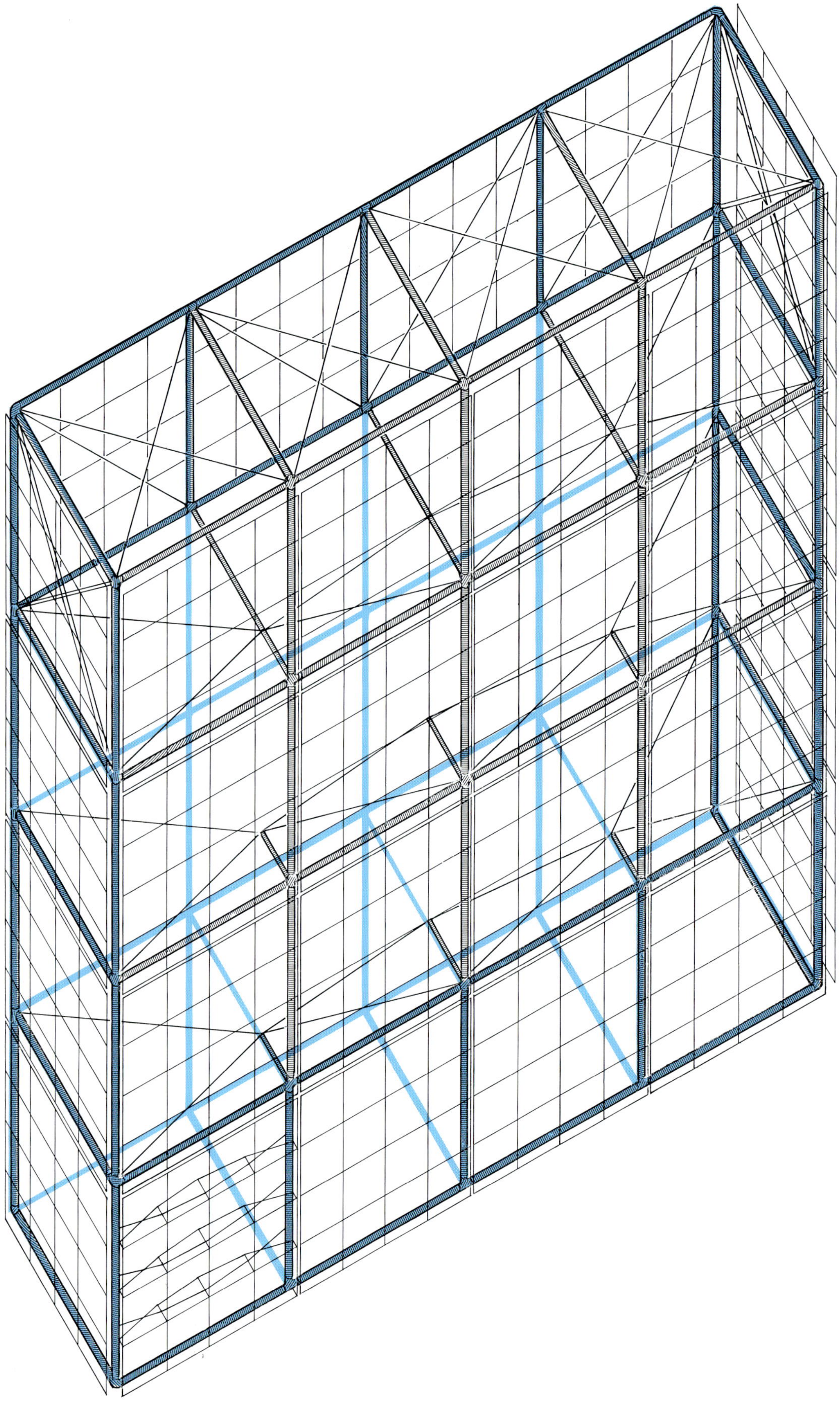

CHAPTER 4

THE STRUCTURE AND OVERALL PERFORMANCE OF A SERRE

The structure of each Serre consists of a welded stainless steel frame that determines the geometry of the whole. As we have seen, this 32 × 32 m frame, subdivided into 8 × 8 m panels, has not changed since the competition in 1980.

Description of the system

The modular geometry of the structure is the same as that of the glass panel grid, but a scale larger. The structure is attached to the main building by two large concrete cylinders clad in stainless steel; these are known as the *piles rondes*. Measuring 24 m in height, they provide the horizontal support that is needed every 8 m at each panel intersection.

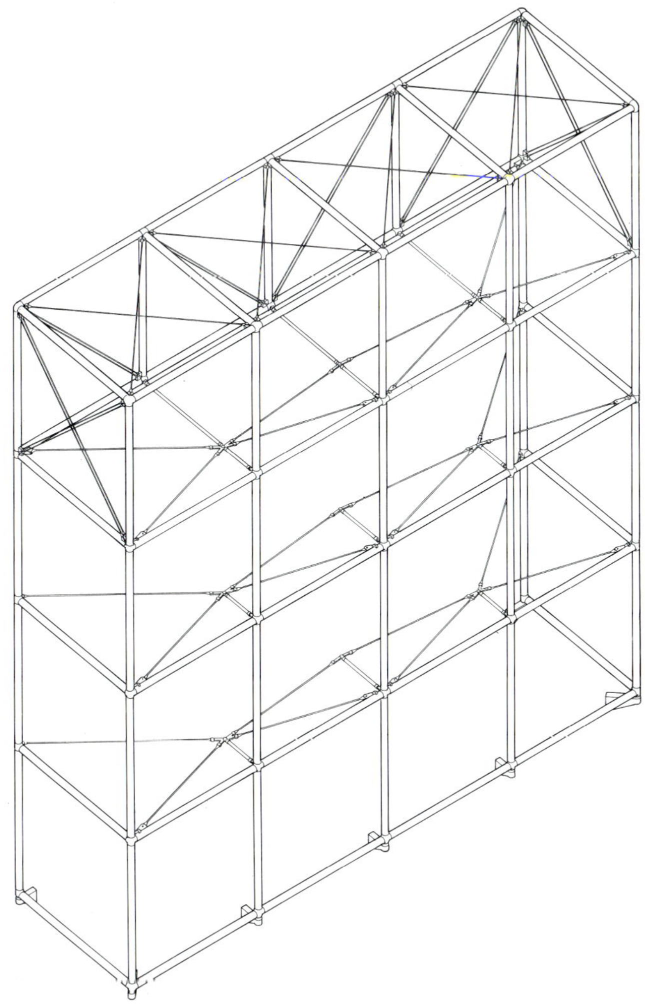

A Serre of 32 x 32 x 8 m is composed of 24 panels of 8 x 8 m.

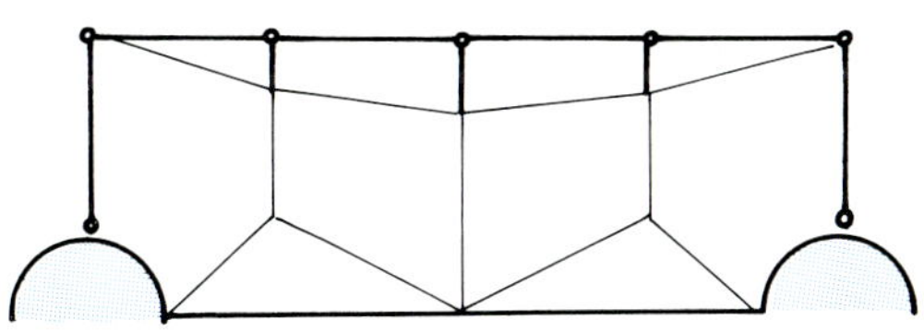

Schematic diagram of the principle of the structure (preliminary project stage).

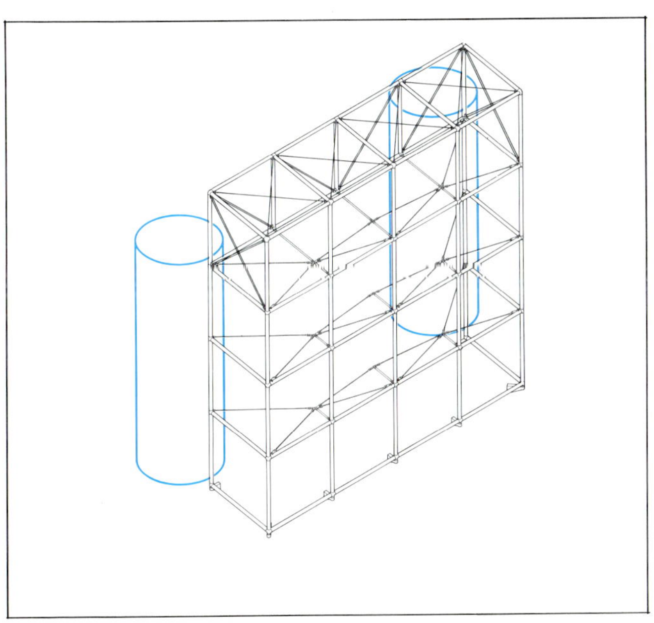

Axonometric of the principle of the structure.

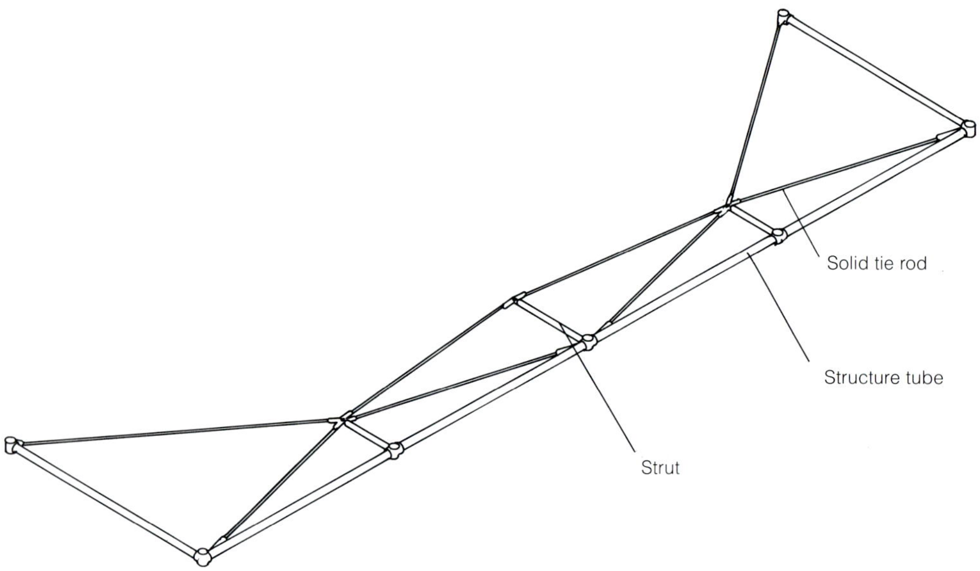

Above:
Principle of the wind bracing of the structure's horizontal mullions.
Right:
Fixing the prestressed solid tie rods to the structure.

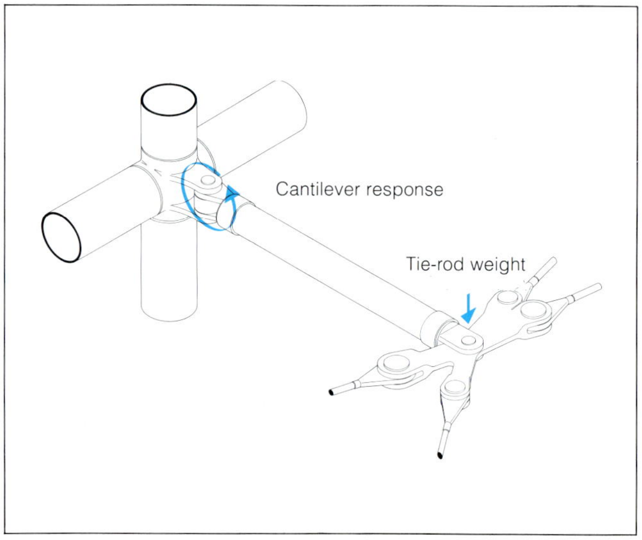

The façade is cantilevered for the height of the last panel, i.e. for 8 m above the *piles rondes*. Thus a rear north-facing aspect is created with a height of 8 m and running for 32 m. Since the Serres are positioned on the south face of the building, each has one face due south of 32 × 32 m, one small upper face to the north of 8 × 32 m, and two faces, one to the east and one to the west, each of 32 × 8 m. Additionally, each Serre is topped with a glazed roof 8 m wide and 32 m long.

The attachments to the *piles rondes* are positioned as follows: at base level (0 m), above the first panel (+8 m), above the second panel (+16 m) and above the third panel (+24 m).

Wind bracing

At each of the first three horizontal levels, the framework of the Serre is secured to the *piles rondes* by a set of prestressed tie rods. This wind bracing reinforces the wind resistance of the horizontal tubes on the south face, as well as on the small north face. It takes the form of two parabolas of solid rods prestressed one against the other, following the same principle as in the cable trusses. The first of these parabolas, curved in towards the building, links the suspension point on the *pile ronde* and the centre of the glazed façade; its main function is to counter outward loads. The other parabola, curved in towards the building, is attached at the end of each horizontal level and penetrates the interior of the Serre to a depth of 3.2 m; its main function is to counter inward loads.

These two tie rods, tensioned against one another, also form a triangulated system that counters loads on the east and west faces. The two tie rods are spaced and fixed to the structure by small tubular members similar to the struts in the cable truss system. The weight of this wind-bracing system is supported by these struts cantilevering from the structure.

The prestress in the tie rods is created in two ways: firstly, by increasing the spacing between the tie rods by means of the struts; secondly, by using a turnbuckle positioned next to the anchor points on the *piles rondes*.

The weight of each façade is supported by seven brackets bolted to the plinth that forms the base of the Serre. This plinth is a concrete structure which holds the earth for the planting. It is inset by 1 m in relation to the horizontal tubes of the steel structure. The steel brackets jut across this gap, cantilevering out above the moat that surrounds the whole building.

The upper part of the Serre, positioned above the *piles rondes*, consists of a panel cross-braced on its east and west faces. This bracing can allow the side panels to function as a cantilever. The cross-bracing is achieved by a double set of tie rods, stretched just enough to stop them from sagging under their own weight. The framework of the roof is equally wind braced in its own plane, which allows it to act as a horizontal wind-bracing beam.

Finally, the rear face (to the north) is triangulated (with single tension diagonals), so that it works as a vertical beam whose span is equal to the distance between the two vertical tubes connecting to the *piles rondes*. It is sufficient only to have single diagonals: the loads they bear are gravitational – whether they be the weight of the structure and glass itself, or snow –

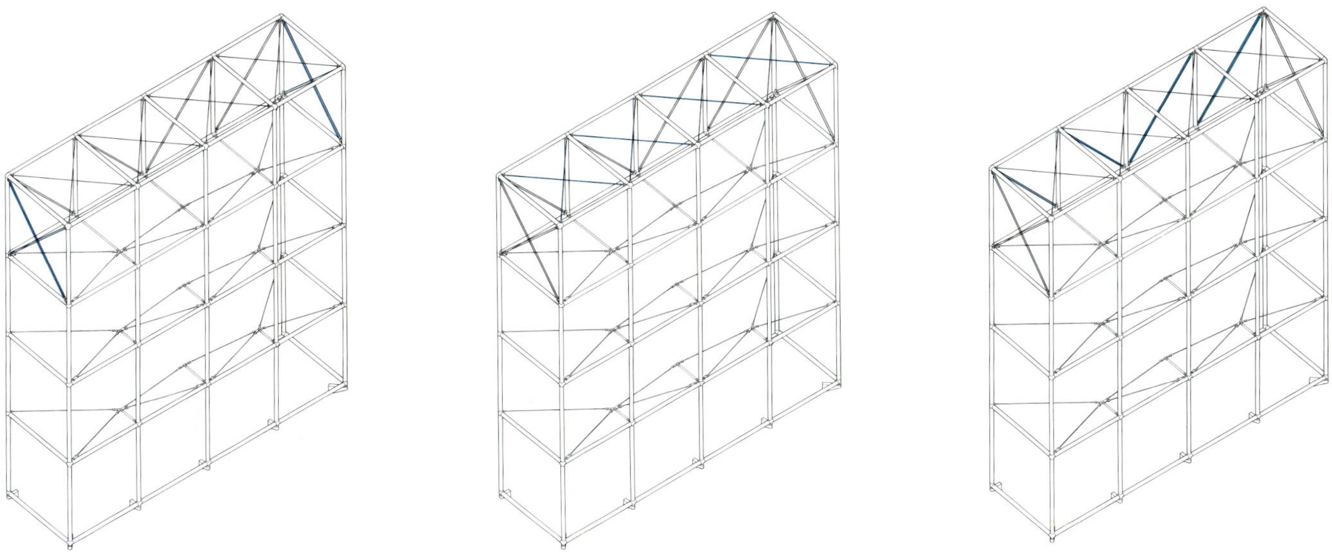

Wind bracing of the upper part of a Serre.
From left to right:
- *on the east and west faces, preventing this structure from sagging under its own weight;*
- *in the roof, tie rods in the horizontal plane: this can act as an exposed beam;*
- *on the rear face, simple diagonals: this is able to act as a vertical beam.*

and the wind loads are too small to cause overall uplift. Consequently, the loads act in a known direction and so the direction of the forces that can be taken by the diagonals is therefore also known. The lower member of this beam, which is independent of the main building, is attached to the horizontal wind bracing at a level of 24 m. This wind bracing carries the wind loads on the back of the façade and stabilizes the beam.

The tubes

The framework of the 8 × 8 m panels is made from stainless steel tubing with an outside diameter of 300 mm. The wall thickness of these tubes varies between 6 and 35 mm (in the most highly stressed conditions). These tubes were fabricated using the technique of centrifugal casting, which allows the outside diameter to be kept constant while varying the inside diameter. Where the tubes cross, they are joined with a cast node. The chosen casting technique allows the aesthetic criteria to be met, while at the same time providing the required strength. The tubes act as simple beams between two nodes. The horizontal tubes bear the weight of the glass as well as their own weight and that of the maintenance equipment. All the cable trusses are attached to the vertical tubes, and they therefore act as vertical beams transferring the loads from the cables to the nodes. They also act as columns transferring the vertical loads onto the brackets at the base of the Serre. The horizontal tubes serve as compression members; with the prestressed tie rods, they form a horizontal wind-bracing system which carries the wind loads back towards the rear of the building. In the same way, the tubes in the upper part of the structure form part of the horizontal wind-bracing beam which has already been described.

The loads taken by the structure

In the design of a building, one starts by evaluating all the loads that it will have to bear; then one continues to the analyses and calculations. Four categories of load have been taken into account for the structure of the Serre: its own weight, environmental loads (wind and snow), forces engendered by temperature variations and the weight of the maintenance equipment (this last is not substantial and did not constitute an important criterion in the design of the structure). We consciously ignored live working loads, which are borne by the planting platform. We did not take into account any other types of load which could be applied to the frame: it would have been too difficult to ensure that the usage of the space in the future would be consistent with the architectural intentions, so we did not design the Serres to have a reserve that would permit additions such as mezzanine floors or lift towers. From an architectural point of view the volume represented by a Serre is, as we have seen, an intermediate space lying between the museum and the park. It was not conceived simply as an exhibition space in which to hang objects, but as a place that would allow visitors to benefit from the park when in the museum and to appreciate the museum when in

Taking the glazing's own weight in the event of breakage of sheets of glass: if the suspension assemblies do not incorporate a spring (in the centre), one of the remaining suspension points is suddenly forced to take a major load; if the suspension assemblies do incorporate springs (on the right), the increase in loads is shared between the two remaining suspension points and the effect of the shock is minimized.

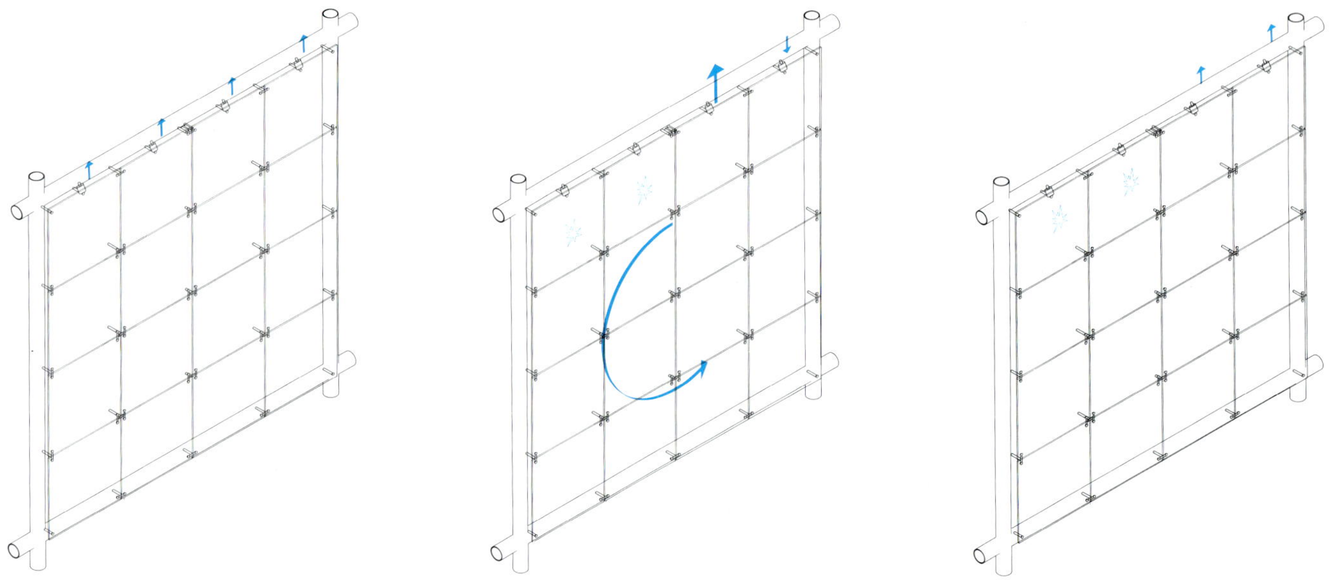

the park. One of the few methods of guaranteeing to achieve this aim is to control the ways the structure of each Serre handles the loads that act on it.

Its own weight

The vertical loads are composed of the weight of the glass itself and that of the structure – the larger of the two obviously being that of the glass. The weight of the structure itself cannot be established until after the structure has been defined. A preliminary estimate is made, based on the form this structure takes from previous experience; thence one can proceed to initial sizing. The loads are then checked and used in all subsequent calculations.

The weight of the glass presents a more complex problem. Once the thickness of the glass has been chosen, its actual weight will be known; and once the glass suspension system has been defined, the bending forces induced in the structure by the offset position of the glass with respect to it can be calculated.

Taking the breakage of glass (described in Chapter 2) into account poses a problem. Following the loss through breakage of a glass sheet, the total weight of the glass is of course less. However, at the same time the loads are redistributed, which could have detrimental consequences for the structure. Moreover, the breakage of a sheet results in a significant temporary overload, since the shock wave is cushioned by distortion and energy absorption in the structure. Because of the difficulty of predicting the effects of such shocks, we introduced prestressed shock-absorbing springs at the glass suspension points as described in Chapter 2. Consequently, the maximum load that the structure has to be able to carry is limited.

Environmental loads

Forces arising from the wind and from snow are the external environmental loads applied to the structure. The statistical aspect of the wind loads was discussed in Chapter 3, with an analysis of the loading and performance criteria for the horizontal trusses. This data remains pertinent in defining the performance of the structure and determining the required level of prestress.

The loads resulting from snow also have a statistical aspect. The maximal weight of snow anticipated for a building such as La Villette is relatively low. As far as the vertical load is concerned, it must be added to the self-weight and to the other vertical loads; the snow load in this sense is not very significant compared with the other vertical loads, but its distribution is unknown: drifting snow can produce highly asymmetric load conditions. It was therefore reasonable to consider a large number of different cases, in addition to that where the maximal weight of snow is spread horizontally over the whole surface. The wind and snow loads, together with the different load patterns that they can induce, are briefly described below.

- Wind loads

Apart from the statistical analysis of wind loads, the appreciation of the different wind-load diagrams represents one of the most difficult problems in the design of the structure. There is little chance of arriving at a symmetrical load distribution on such a large surface and in an urban environment, where numerous obstacles lying in the path of the wind can create turbulence. Significant positive and negative pressures can occur near the corners of the Serre. It is advisable also to take into account forces parallel to the building and combine them with forces perpendicular to the building, so the wind can be trapped between two Serres and exert a lateral force on each of them, while inward or outward pressures are applied to the whole of the Serre. The diagrams (right) show the most important possible effects which must be taken into account for the Serres.

The calculation of the wind loads requires an important interactive analysis. What is needed is to select the wind forces that are liable to produce the largest forces on individual structural members. This necessitates a preliminary study of the structure in order to become familiar with its behaviour under different load conditions; when this information has been obtained the load patterns most likely to be critical are selected. In the case of a structure such as the Serre, this interactive analysis needs to be carried out very carefully.

• Snow loads

The snow load only affects the roofing and the north façade. Given that the roof is a simple beam that spans from front to back, the actual distribution of the loads is not particularly significant: the maximum overall load will always result in maximum forces in the structure. Thus, drifting of the snow on the roof is not a problem. However, an accumulation of snow against the glass wall can create forces in the north face; this constitutes a case of a particular pattern that must be considered in the calculations.

Effects of temperature

The main temperature-induced stresses are due to differential heating of the structure of the Serre in relation to the main building which supports it. Each horizontal wind-braced element of the structure is linked at its ends to the *piles rondes*. It is assumed that the distance between these elements does not change because each *pile ronde* is separately anchored into the ground. The structure of the Serre can heat up in summer as a result of solar gain and be cool in winter; this will induce stresses in the structure and a diminution in tension in the prestressed elements. These stresses arising from changes in temperature must be

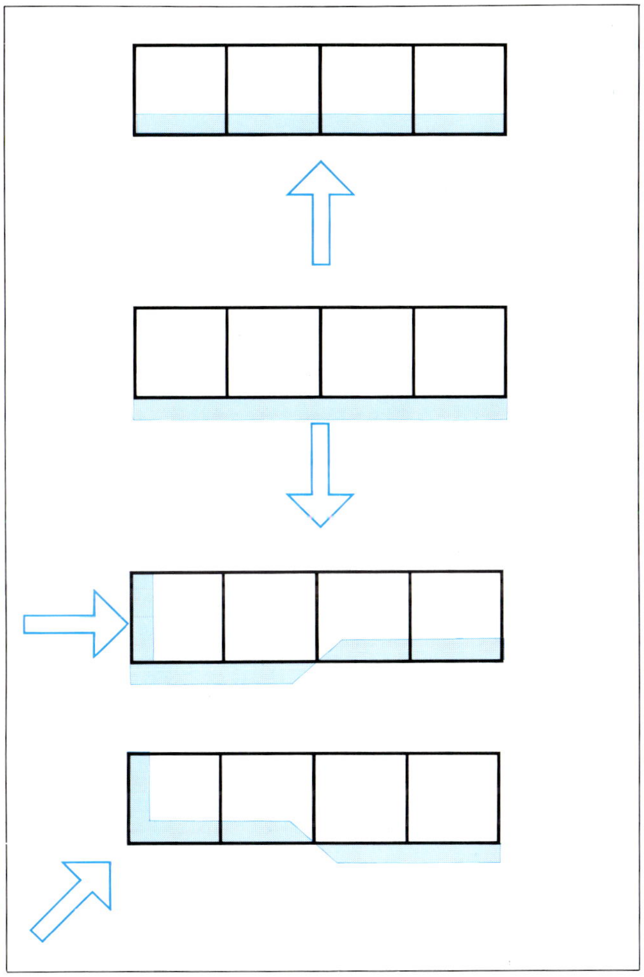

Wind effects taken into account in the calculation of the Serres. From top to bottom: positive pressure; negative pressure; local combined effects of positive and negative pressure.

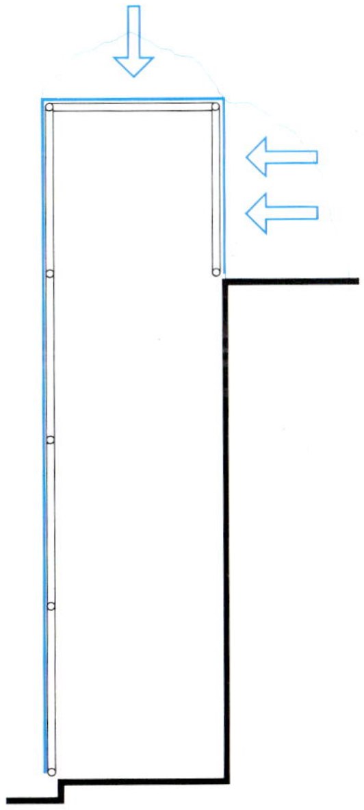

The effects of snow loads on a Serre.

considered where they are likely to result in major alterations in the forces in the system. The variation in temperature used in the calculations was from +35°C to –20°C, implying a variation in the height of the façade of ±9 mm. Furthermore, this means that the façade can grow or shrink vertically at the attachment points to the *piles rondes*; thus, hinged connections were required.

How the structure works

As explained above, there are three types of load that must be considered in order to analyse how a Serre works structurally: vertical loads, which include the weight of the Serre, loads due to snow and loads due to maintenance; horizontal loads from wind; and the effects of temperature. These last entail certain modifications to the constraints; it is therefore reasonable to establish a series of hypotheses relating to the tolerances to which the structure has been erected and prestressed and to analyse the resulting implications. Let us now explain the reaction of the structure to different load situations.

How the structure carries vertical loads

Each face of a Serre transmits loads downwards individually. This is done in mostly the same way for each of them. As explained, the tubular members in the roof act as beams on two supports in transferring vertical loads.

The vertical loads transmitted by the horizontal elements in the south, east and west faces create a moment on the vertical tubular members of the structure; primarily due to the weight of the glass, this moment is located outside the plane of the structure, which creates torsion and flexion of the tubular members in it.

The vertical columns transfer these loads onto the brackets at the base of the Serre.

It is only the framework of the north face which is different. It behaves as a beam 8 m high, with a span equal to the length of the back face, which transfers the loads onto the tubular members next to the *piles rondes*. The central three sets of vertical bars are an integral part of this beam.

An important feature of these vertical loads lies in the fact that once the structure is completed, they change very little. No distortion is therefore transmitted in the plane of the glazing. This glazed plane is flexible as long as the joints remain unsealed; but once the joints are made, the silicone resists deformation so that the whole of an 8 x 8 m panel behaves as a rigid sheet. Should a major deformation occur (because of the glass breaking or for another reason) the prestressed springs, which have a fuse-like action, come into play.

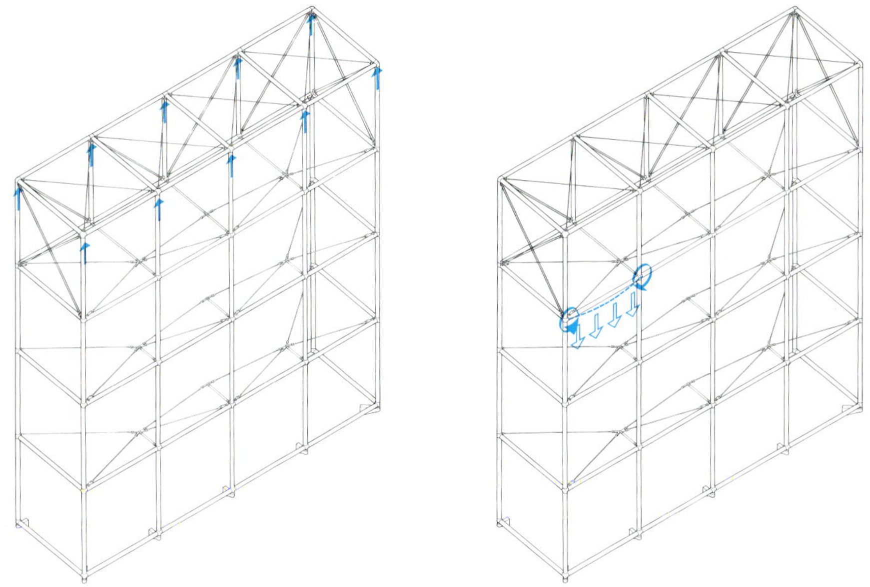

Bearing vertical loads:
- *left, the tubes in the roof transmit vertical loads like beams onto the top of each of the vertical tubes of the structure;*
- *right, the moment created on the vertical tubes of the structure, due to the vertical load transfer by the horizontal mullions of the structure, causes bending in its nodes.*

How the structure carries horizontal wind loads

The horizontal loads can occur in different patterns, and resistance to them is provided at each main level by the U-shaped horizontal frames and by the prestressed tie-rod system that works with them. Thus the wind-bracing system comprises three parts: the U-shaped frame (a), the prestressed tie rods acting against the inward loads (b), and the prestressed tie-rod system taking the outward loads (c); (b) and (c) are prestressed against each other. They work together until the tension lost in the system causes a member to go into compression.

For example, for a load coming from the outside, i.e. a positive wind pressure, (b) offers resistance through a growing tension and (c) through a decreasing tension until it reaches zero; the tie rod (b) is then left to offer resistance on its own. As with the cable truss (see Chapter 3), the prestressing serves to rigidify the system in the lower load ranges – statistically the most likely. Prestressing the tie rod (b) results in compression in the front horizontal member of the façade, compression which grows with the increase in tension in (b).

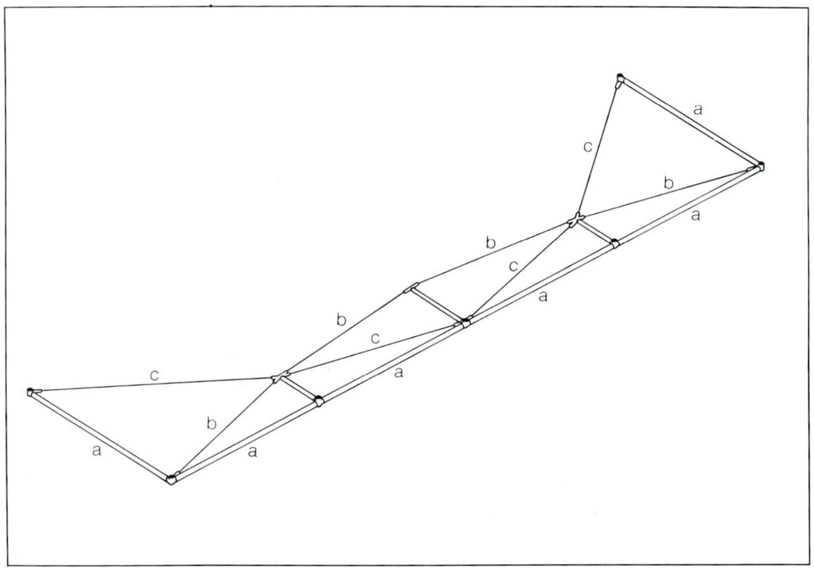
a
c
b
a
b
c
a
b
c
c
a
b
a
a

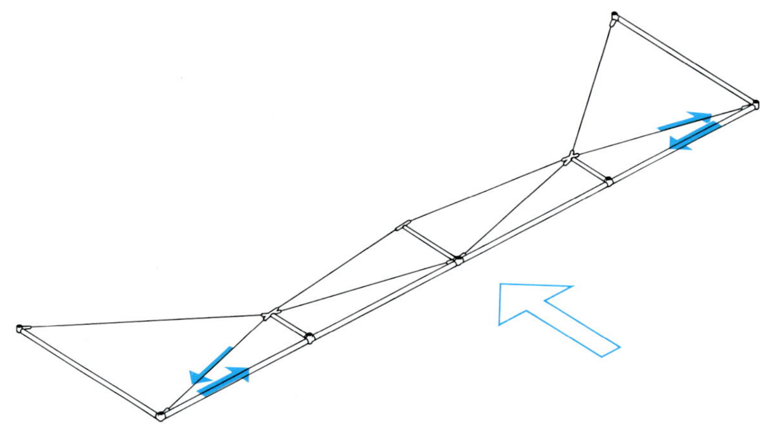

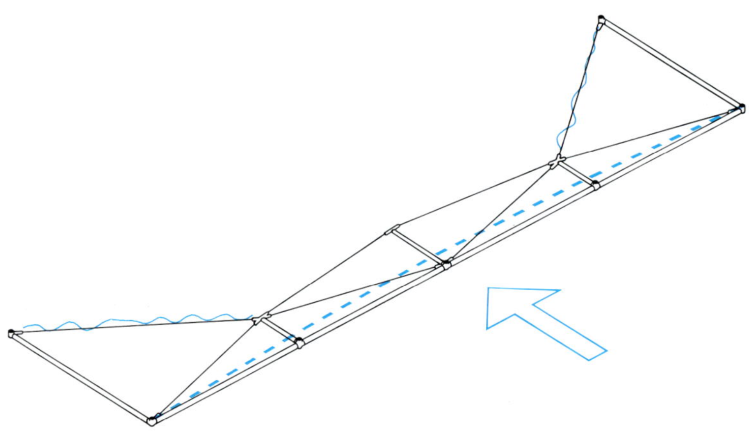

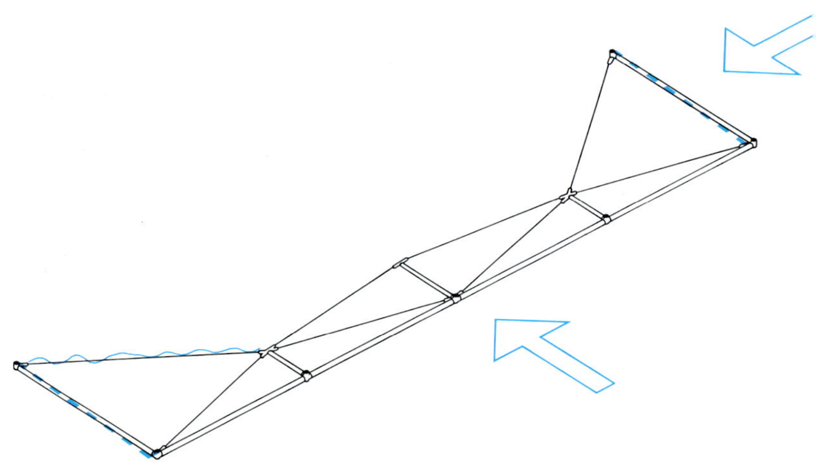

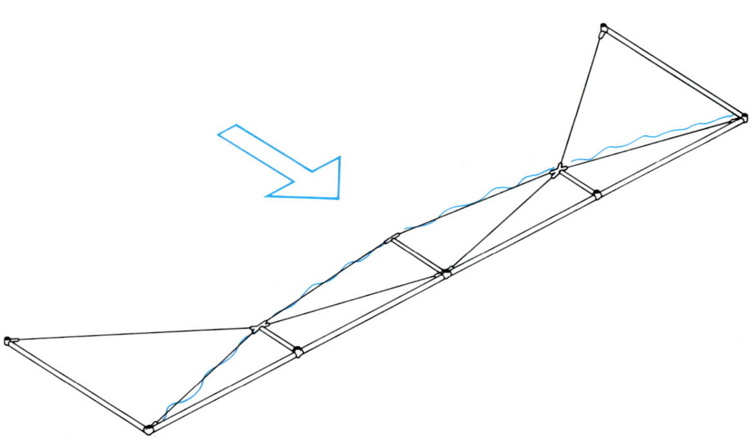

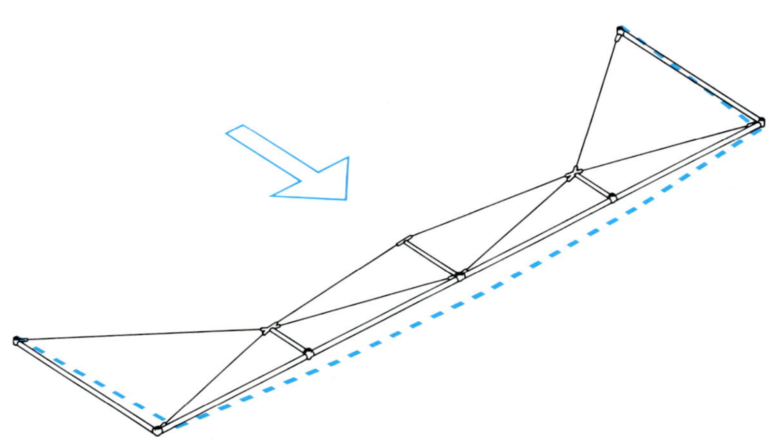

Opposite:
Left, the wind bracing of the structure is ensured by three elements: the U-shaped tubular frame (a), the tie rods taking the exterior loads (b) and the tie rods taking the interior loads (c).
Right, the wind bracing's principle of action.

This system makes full use of modern computer techniques. The structure and bar-bracing system changes its topology (structural form) according to the load conditions. Modern structural analysis allows for changes in the form of a structural system in the course of the calculation. This implies exclusive use of tension members (because they must 'disappear' from the structure when they go into compression). In this topologically free structure two special cases can arise.

The first arises when a large inward pressure is accompanied by lateral pressure on the east or west face. Because of the inward pressure, tie rod (b) can 'lose' its tension; but it still remains indispensable in resisting lateral loads. To bring this member into play once more, a lateral displacement of the structure occurs. Predicting this type of behaviour would have been difficult using classical analytical techniques. Such techniques assume in effect that all deformations are small in magnitude with respect to the geometry of the whole system and are consequently negligible, and, further, that all elements contributing to the structural integrity remain active under all load conditions.

The second special case also occurs when major inward loads induce large tensions in tie rods (b) and make tie rods (c) go slack. This involves a large compression in the horizontal tubular members (a) of the structure. This means that the front horizontal tubular member could buckle if it were poorly restrained at the nodes. Buckling is a phenomenon that occurs in compression members: the axial force gives rise to extra internal loads, which at some level of axial force overcome the stiffness of the member.

The degree of restraint at the intermediate nodes is fundamental in determining the force at which buckling will occur.

To examine this problem the front tubular member was loaded to failure and this load was compared with that which a classical calculation of buckling would predict: the two values were very close. The stiffness of the jointing required to satisfy the predictions of classical theory was in fact higher than was found to be necessary in reality.

Two particular instances of the wind bracing working:
- *above, with an inward pressure and lateral measure: the tie rods (c) 'lose' in tension under the first, but not totally in order to resist the second;*
- *below, with an inward pressure: the tie rods (c) 'lose' in tension; only the resistance of the nodes prevents buckling.*

How the structure reacts to temperature changes

As has been explained, temperature changes can modify the actual prestress in the members compared with the intended levels. All the calculations were made with different hypotheses of prestressing, while taking into account temperature differentials between the façade and the building ranging from +35°C to −20°C, and allowing for margins of error resulting from constructional inaccuracy. These hypotheses do not affect the general strength of the building, but mainly apply to the rigidity of the structure, which we already know is not a critical effect.

Hierarchy of the structure

The structural functions of each element can be defined in terms of the structural hierarchy of the system, of which the main frame structure is the primary element. The wind-bracing systems, the attachments to the building, the glazed assemblies and the bracing trusses are all linked to the primary structure by castings welded to the tubular members or by lugs forming part of the main structural frame cast nodes.

All the secondary elements of the structure are stretched and articulated; their function is to reinforce the primary structure. In visual terms they are lighter than the elements of the structure to which they are fixed. Their attachment to these is achieved by what might be termed 'sculpted' castings in such a way as to transfer loads into the structure while avoiding excessive local reinforcement. These secondary components have to adopt specific forms in order to satisfy the load transfer requirements. The fact that these elements are castings, accentuates the jointed nature of the assemblies and forms a contrast with the essential character, giving the impression that the structure is unbroken and continuous. Certain elements lower down the hierarchy are also attached to the structure by special castings: the V-brackets, the spring connections, the perimeter glass suspension points and the maintenance rail.

The visual and functional separation between the elements constitutes an important conceptual characteristic in the legibility of the structure. As far as possible, the project tries to show the way in which the loads are transferred into the different parts of the structure and to separate the individual functions of each element to make the way in which each assembly works understandable. This contributes to making the predictability of the performance of the system easier – a critical point when dealing with a material as fragile as glass. The articulations and spherical bearings incorporated in the links between the elements guarantee achievement of this objective of understanding the structural behaviour.

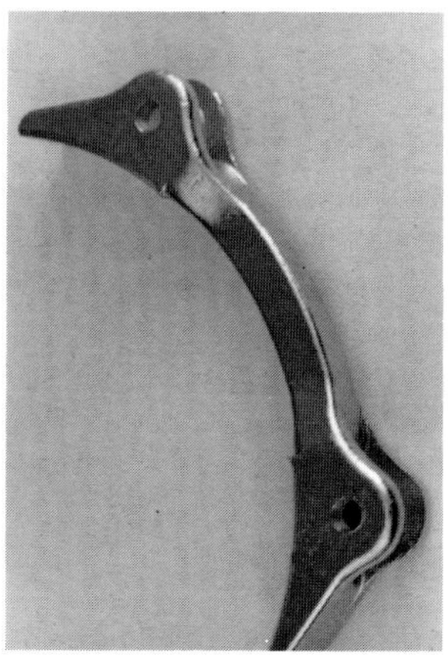

Top:
The fixing lugs on the wind bracing tie rods of the beam.
Middle:
The glazing's perimeter suspension points.
Bottom:
Component for fixing the sprung suspension assemblies to the framework.
Right:
Component for fixing the solid wind bracing rods to the structure.

The materials and manufacturing methods of the structural elements

The principal components of the structure of the Serres are: the framework with its tubular members and nodes, the trusses, and the wind-bracing system (solid rods with end fittings and cast attachment points). The different procedures used to make these components will now be described briefly.

Two fundamental considerations had to be addressed from the beginning: the surface finishes and the quality of the materials. There are various surface treatments available for stainless steel (the material from which all the structural elements are made). The final quality that can be achieved depends on the composition of the material, on the way in which it was manufactured and on the polishing method used. This choice may be complex, but it is important because the material's resistance to corrosion will depend on it.

Stainless steel

All the metallic components making up the Serres are made from stainless steel, which differs from standard steels in that it contains chromium. There are three types: martensitic, ferritic and austenitic; these are defined by the chemical composition of the alloy and by the heat treatment the material has undergone.

Most of the steel used in the Serres at La Villette is austenitic stainless steel – characterized by a nickel content of around 80%. Austenitic stainless steels lend themselves easily to being worked and welded. Additionally they offer a higher level of resistance to atmospheric corrosion than other steels: the presence of molybdenum in the alloy improves not only its visual qualities, but also increases its passivity (or chemical resistance) to corrosive elements in the atmosphere.

The mechanical strength of the steels used in the building is defined by their elastic limit (σ_e). Among ordinary steels, the value of σ_e will be found to lie within the range 240 to 360 MPa, whereas the elastic limit for the austenitic stainless steel used in the Serres was 205 MPa. Other specialized stainless steels with a higher mechanical strength are used in certain components, notably for the axes of the spherical bearings in the glass suspension points. By virtue of a specific toughening process and a special chemical composition, σ_e in this case attains a value of 900 MPa.

The attachments to the piles rondes

These are steel assemblies made to measure after the *piles rondes* had been cast and exactly surveyed dimensionally. A particular tolerance was allowed in these details, given the differences in expansion and contraction between the metal framework and the primary structure in concrete, due to differences in temperature and the coefficient of thermal expansion. These assemblies also allowed the system a certain rotational freedom.

The nodes

The nodes were made in ceramic moulds. Experience gained from the Pompidou Centre played a large role in the decision to use cast pieces. The process consists of first making a positive model of the item (often in wood) and from that a negative mould in ceramic.

Re-entrant shapes or those that cannot be removed from the mould once the piece has been cast are made in polystyrene or wax: they can be burnt out or melted after the casting. Each part of the mould was marginally oversized to allow for shrinkage on cooling.

The molten metal is poured into the mould through a funnel-shaped *masselotte*; this holds a sufficient weight of metal to force the molten metal into all the cavities and then retains all the impurities in the metal that floats to the surface. It is removed once the piece has been cast, leaving a 'scar'. The position of this

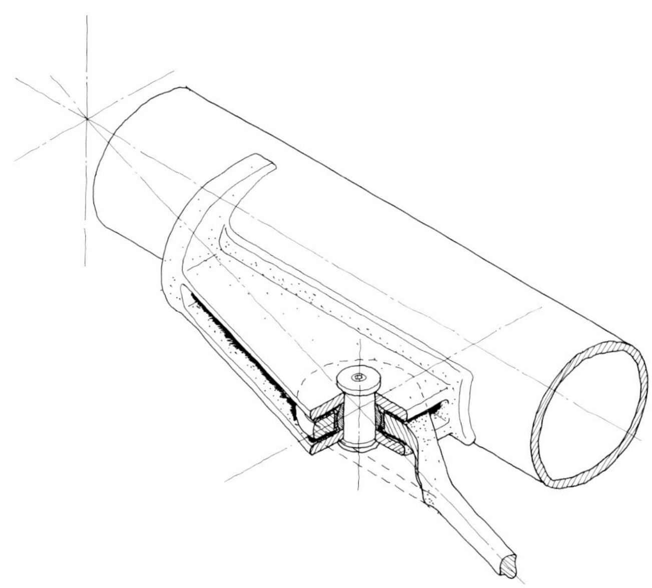

Top:
A node from the structure.
Bottom:
Electrolytic polishing treatment (acid bath) of a stainless steel component.

scar is important since it can affect the uniformity of surface treatment. The cast piece is cleaned by blasting with fine grit and is machined at the contact points with the tubular members or at other connections requiring tighter tolerances than those which the casting process offers. It then undergoes an electrolytic polishing process to eliminate any surface imperfections and to give it a shiny finish. The final result is an orange-peel appearance, slightly roughened, and very different from any texture produced by other finishing processes.

The technique of electropolishing involves submerging the piece in a bath of acid, one electrode being connected to the piece and the other to the side of the bath. The current passes from the cast piece to the bath and eats away the surface of the piece, revealing the pure untarnished metal underneath.

Certain problems in homogeneity arose when this finishing process was first used: the thickest pieces came out very matt, while the thinner extremities were very shiny. These problems at first caused us considerable concern, but they were resolved by very accurate adjustment of the current passing through the piece and the time it was left in the bath.

The tubes

The tubes between the nodes in the framework were made by centrifugal casting. This technique was used because the required thickness of the tubes excluded the more usual welding method (where a flat strip is bent to form a tube and the join is welded along the entire length). Centrifugal casting allows a variation in thickness in the tube wall without changing the mould dimension, provided of course that the external diameter remains fixed. In this way a considerable economy in material is achievable.

The wall thickness of the tubes is varied according to how much work they are expected to perform. For example, the corner tubes, which permanently take the tensions from the cable trusses, are 35 mm thick, whereas the horizontal tubes at the base of the façade have a thickness of only 6 mm.

The tubes are cast using a mould 2 m long which turns at high speed on its own axis while the molten metal is poured in; centrifugal force spreads the metal evenly to form the walls of the tube. Different thicknesses can be achieved by varying the amount of metal poured into the mould.

The sections, with a maximum length of 2 m as a result of the way they are made, are then butt welded. These welds introduce local stresses in the tube,

Attachment to a pile ronde.

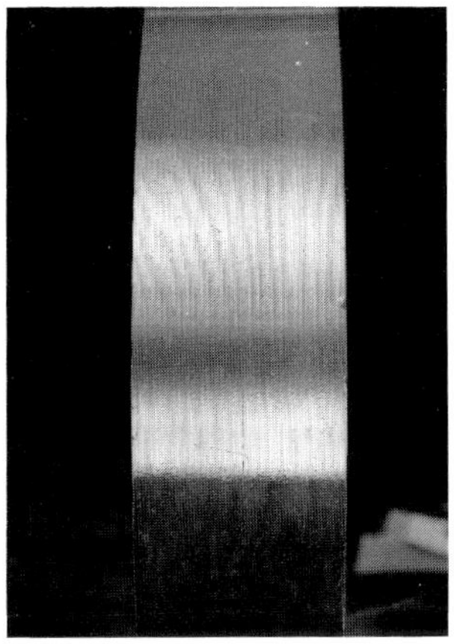

The finely striated appearance of a tube after turning.

which are then relieved by further heat treatment (called 'annealing') of the complete welded item. Finally the tubes are machined along their whole length to yield a uniform finish.

Finer machining followed by band polishing allows a very smooth finish to be achieved. The technique of sand blasting or blasting with glass beads can give a satin finish. The manufacturers refer with pride to the stainless steel rollers used in glass production as an example of the degree of perfection achieved in their tubes.

The tube is machined on a lathe (a pointed tool passes slowly from one end to the other of the spinning tube to remove the superficial layer of metal). This operation leaves the tube with a surface that is finely striated. The appearance of these striations varies with the dimension of the point of the tool and the number of passes along the tube.

In the sample shown opposite, the machining marks are fairly deep. The roundness of the tube is strongly expressed: the grooves from the turning catch the light well. A relatively roughened surface finish was sought so that the finish on the tubes would blend with the orange-peel appearance of the nodes' surface.

Assembly of nodes/tubes

The cast nodes and the centrifugally cast tubes were welded. No attempt was made to disguise the welds with cosmetic treatments: they delineate the boundary between the node (with the orange-peel appearance) and the tube (with the roughened finish). The different fabrication techniques are thus clearly expressed.

Three types of welding were employed: (1) certain horizontal tubes were factory welded (the tube rotates against a fixed electrode); (2) other horizontal tubes were welded on site (the tube remains stationary while the electrode moves, giving a circular weld which is very difficult to achieve because parts of it are done by hand); (3) the vertical tubes were welded on site (fixed tube, moving electrode). In this last case, only small fillets of weld could be applied because gravity tends to deform larger fillets while the metal is still liquid.

Each of these three types of weld has its own appearance: the horizontal factory welds are very smooth and even, the on-site horizontal welds are much rougher and their finish includes slightly irregular grooves, and the vertical on-site welds have the appearance of bands of string one above the other.

Wind bracing

The tie rods are made from solid rods with end pieces incorporating spherical bearings at each end. They are joined through cast pieces, either to each other, or to the structure. The tie rods and their forged end pieces were specially made for the Serres – spherical bearings are incorporated in the ends to eliminate any bending moment resulting from their own weight. Cast connections were used at the points where the tie rods are fixed to the tubular structure. It was important to express the articulation between the

Right:
Detail of a node–tube assembly: here, weld type 3 (horizontal), weld type 2 (left), weld type 1 (right).
Far right:
From top to bottom, welds of type 1 (factory) and types 2 and 3 (on site).

A link node between the tubes in the structure.

tie rods and the structure to reveal the hierarchy in the system. Excluding the nodes and other cast pieces for the glass suspension system, 15 different types of casting were used. These are mostly female components with two lugs welded to the tube and their male counterpart welded to the ends of the wind-bracing rods. These butt-welded joints had to be checked individually by X-ray to detect possible internal weld defects; each weld between a rod and its end pieces was test loaded to 90% of its load-bearing capacity.

A considerable dimensional rigour was imposed on the contractor, since only a small number of the components of the structure were adjustable. Tubes and rods had to be precisely cut to length having allowed for weld shrinkage. These requirements ensured that all the dimensional issues were addressed in the controlled environment of the factory and not on site. The only adjustable elements are the stub struts of the horizontal trusses: their ends are threaded – in other words, by turning the strut it increased or decreased in length – thereby enabling a final adjustment to the façade's planarity.

Erection

The design of architectural projects frequently omits consideration of the questions of transportation and erection of the component structures. These, however, have a far from negligible importance for the structure and its appearance.

The structure of the Serres at La Villette relied, as far as possible, on prefabrication. The 8 m tubes are first assembled two by two to yield tubular elements 16 m long – the longest the transport jig frames would allow. All the details for the glass suspension and the attachment of the tie rods and cable trusses were pre-welded to the 16 m tubular elements. These assemblies were then transported to the site by lorry.

The welding of the 16 m tubular elements, as well as of the 8 m perpendicular tubes, was carried out in position. These welds had to be effected with a high level of precision to allow for weld shrinkage and the prestress in the trusses.

The first three levels were assembled and provisionally braced with scaffolding. The upper level (a parallelepiped measuring 8 × 8 × 32 m) was pre-assembled on the ground, then hoisted into position in one single piece.

The final object constitutes a work that is internally consistent, yet in which each constructional element is clearly and individually legible.

The connecting strut from the solid rod wind-bracing system to the structure. At the bottom left a sprung suspension assembly can be seen and at the bottom right a one-point assembly (suspension of the glass at the perimeter of a panel).

Putting the structure of a Serre together.
The 16 m long pre-assembled tubular elements are delivered on site. The first three levels are progressively welded and provisionally braced with scaffolding. The last level is entirely pre-assembled at ground level and lifted into place as a single entity. Then the cable trusses are installed (Chapter 3, p. 77) and the glazed assemblies are put in place (Chapter 1, p. 41 and Chapter 2, p. 61).

Principles of a Serre's climate control.

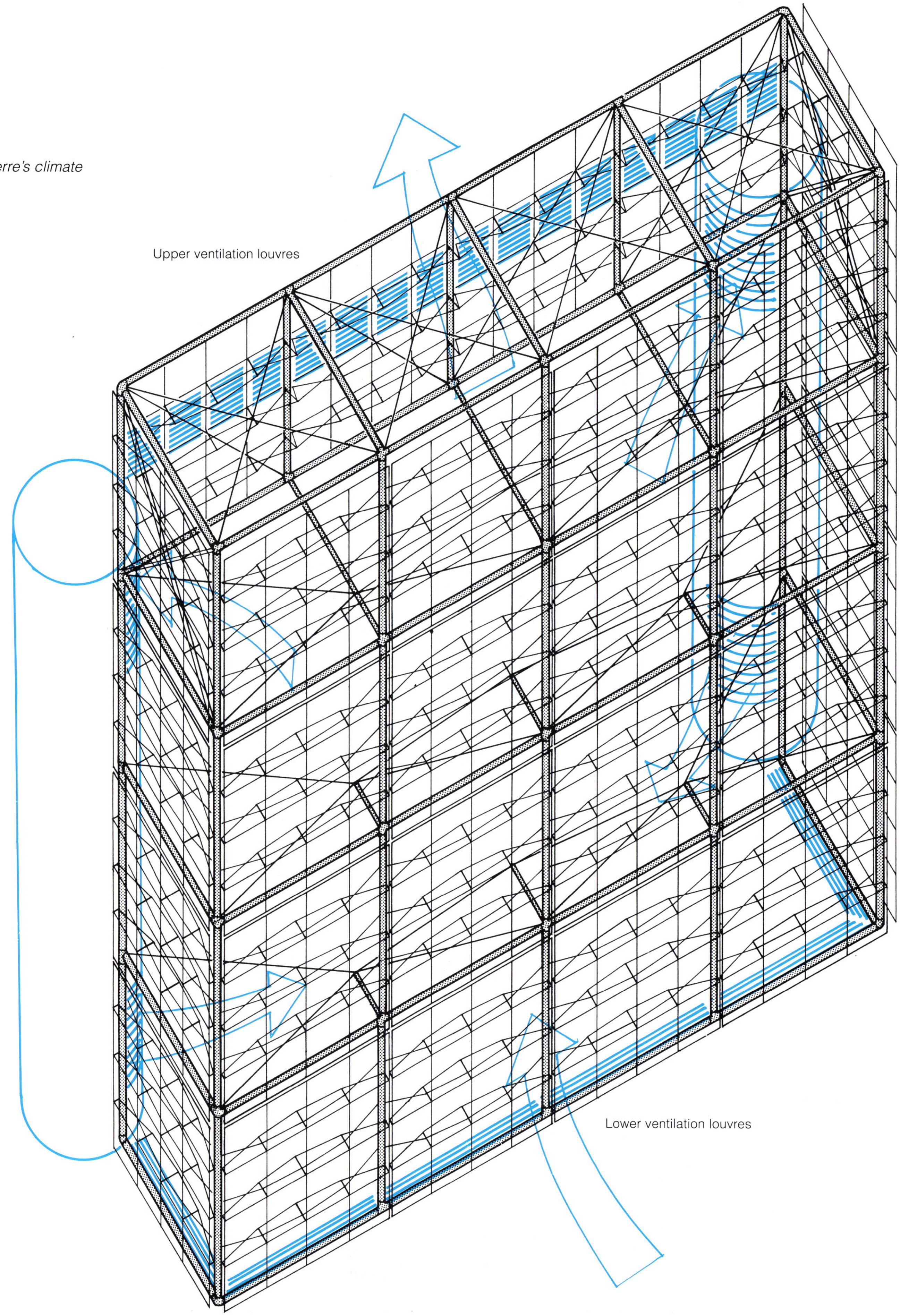

THE CLIMATE INSIDE THE SERRES

The Serre housing the bamboos.

The theme of Adrien Fainsilber's competition project, 'the relationship with nature', must be remembered when discussing the climatic qualities of the Serres.

The Serres play a key role in the architecture of the project as a buffer zone between the museum and the park. In a didactic manner they express technology on the park elevation, while allowing the foliage and greenery of the park to enter the museum. The Serres must provide an environment in which plants can survive. Furthermore, the solar energy trapped by the large glazed volumes must be exploited. For this part of the project, consultation with climate specialists for the building industry was necessary. The study was conducted jointly by SGTE and Novelerg.

The basic principle of this research was to limit any possible extremes of temperature, both for the sake of the plants and to recover the solar energy.

Interior temperatures

It is anticipated that the temperature inside the Serres will be controlled all the year around: therefore a compromise between minimal energy cost, maximal solar energy recovery and finding a favourable environment to encourage plant growth had to be found.

The following conditions and principles for temperature control were selected:

- winter (for an outside temperature of –7°C) (hot air heating giving the required temperature). $\theta_i > 5°C$
- mid-season (heating off) $5°C < \theta_i < 13°C$
- mid-season (energy recovery) $13°C < \theta_i$
- summer (for an outside temperature of +32°C) (static ventilation) $\theta_i < 35°C$.

The use of solar energy

Two working hypotheses have been developed.

1. To give energy autonomy to the façade and hence to create an energy buffer on the south façade for the museum (main building). The climate inside the Serres has been computer simulated for the different seasons. The study shows that the shape of the Serres and the contained volume of air create a kind of microclimate. The upper 8 m bay cantilevers up above the terrace roof and consequently the air inside cools faster. This 'cold head' phenomenon produces a natural air convection inside the air volume. The solar energy trapped by the Serres is stored in the base and used when needed.

2. To use the energy recovered in the Serres according to the environmental needs of the museum. The energy produced in the Serres is recovered by heat pumps located in the museum's refrigeration and heating plant. An energy saving of approximately 220 to 240 MWh/year is possible, which is 3.5% of the museum's needs.

The second hypothesis was finally selected. However, the computer simulation studies conducted during the research work undertaken on the first hypothesis were useful for understanding the microclimate behaviour.

Climates encouraging plant growth

Four factors must be considered to ensure the well being and growth of the plants:

1. humidity must remain above 60%;

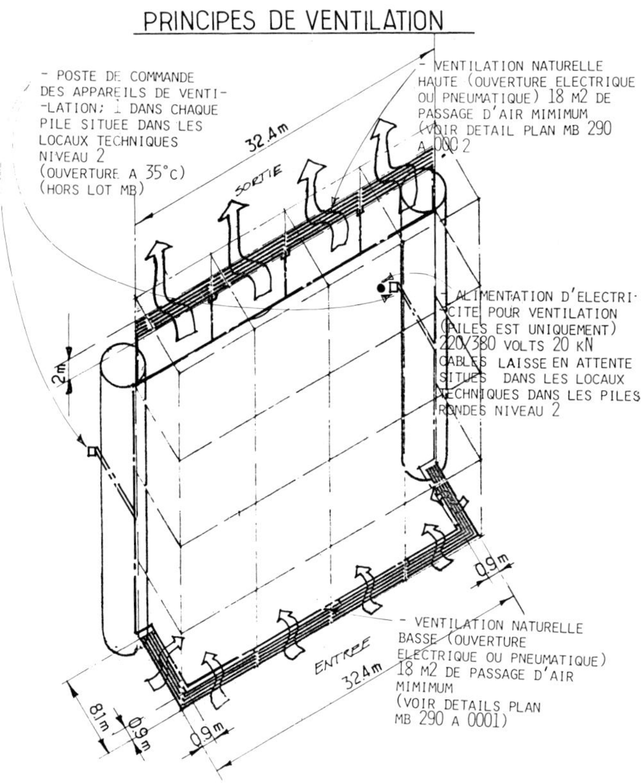

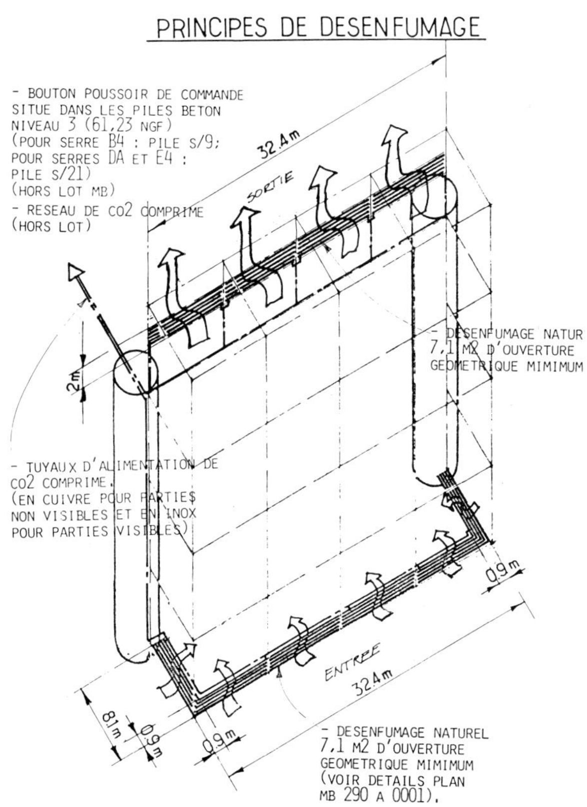

2. there must not be any marked climatic changes;
3. solar radiation must be well controlled;
4. the air must be renewed at the rate of three volumes per hour.

Climate control equipment

There are two parts to this:

– the equipment related to the infrastructure and the building's envelope;
– equipment changing according to the plants' needs.

Inside the museum a refrigeration and heating plant is linked to the Serres by ventilation ducts travelling through the museum. Ventilation grilles incorporated into the *piles rondes* provide a hot air flow from the lower level of each Serre and an air intake at the upper level. Apart from this central heating and extraction system, variable louvre vents are incorporated into the glazing of the Serres.

At the base of the Serres, within the floor section cantilevered over the moat between the supporting brackets, openings are provided which correspond to those made in the upper part of the north face. These small vents are controlled automatically by the central system.

Each group of plants requires its own climate control system (control of humidity, temperature, watering and sunshine). Let us look, for instance, at the system (researched in cooperation with the landscape gardener Kathryn Gustafson) developed for the bamboos in Serre no. 2.

The humidity is measured by probes located at various strategic locations inside the Serre. The data are sent to the refrigeration and heating plant, where corrections are made.

The temperature is also measured by probes located inside the Serre. The data obtained regulate the setting of the small vents and the change in the air temperature coming from the refrigeration plant.

The plants are automatically watered by poles fitted with mist spraying nozzles and by direct watering in the ground.

The bamboos do not require solar control. Nevertheless, blinds can be fitted to the structure, and the glazing suspension rods are extended at their ends to allow the hanging of automatic rollers.

Since each plantation scheme needs a climatic control system to suit it, the fluids, air and electricity services are laid out so as to provide maximum flexibility.

Above:
The upper ventilation louvres positioned on the rear face of the top level of a Serre.
Left:
Principles of the ventilation system.
Opposite:
Working drawings of the roof.

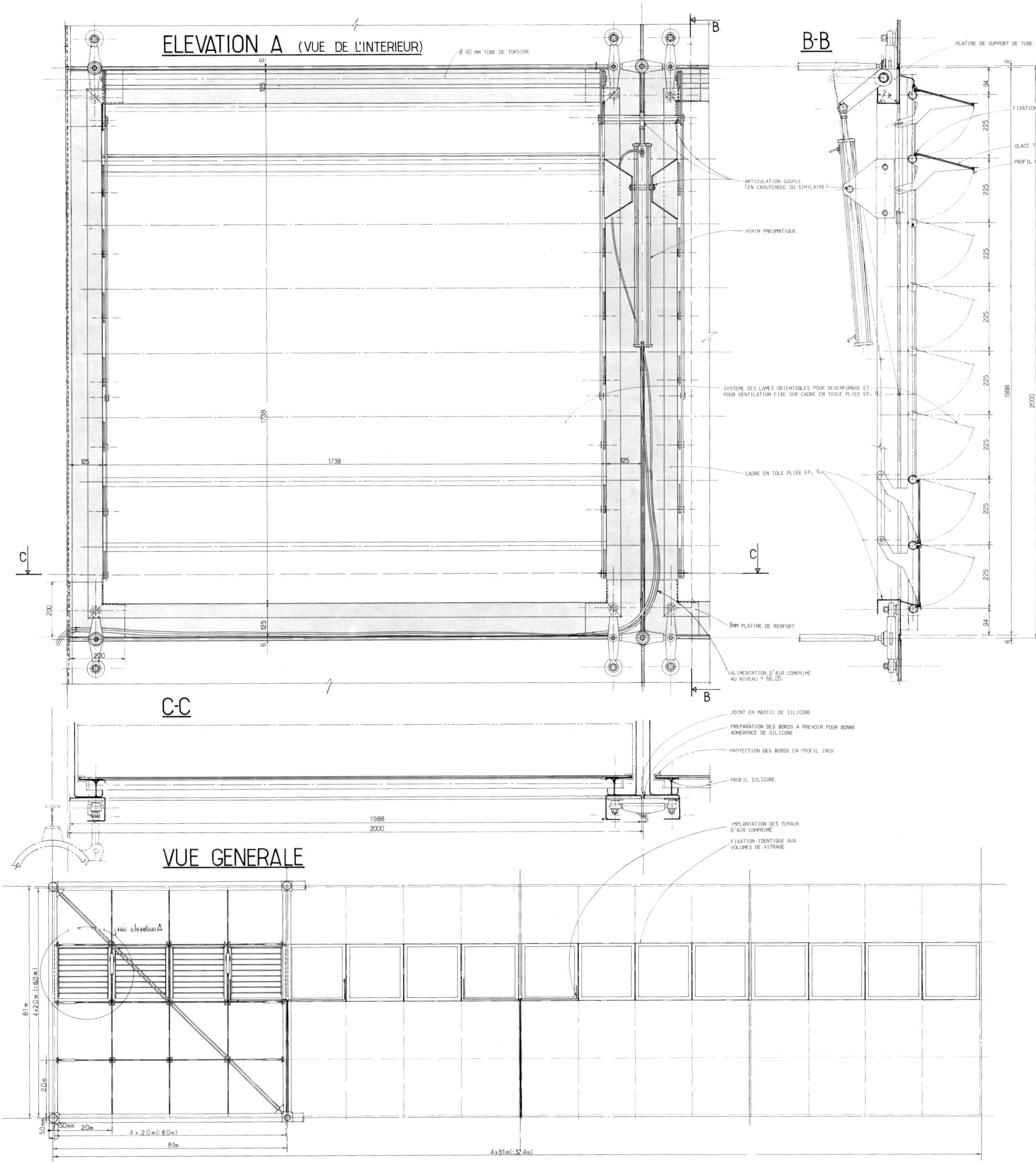
ELEVATION A (VUE DE L'INTERIEUR)
B-B
C-C
VUE GENERALE
Ø 40 MM TUBE DE TORSION
PLATINE DE SUPPORT DE TUBE DE TORSION
FIXATION DES LAMES DE VERRE
GLACE TREMPEE
PROFIL EN SILICONE
ARTICULATION SOUPLE (EN CAOUTCHOUC OU SIMILAIRE)
VERIN PNEUMATIQUE
SYSTEME DES LAMES ORIENTABLES POUR DESENFUMAGE ET POUR VENTILATION FIXE SUR CADRE EN TOILE PLIEE EP. 4
CADRE EN TOLE PLIEE EP. 4
8MM PLATINE DE RENFORT
ALIMENTATION D'AIR COMPRIMÉ AU NIVEAU + 66.05
JOINT EN MASTIC DE SILICONE
PREPARATION DES BORDS A PREVOIR POUR BONNE ADHERENCE DE SILICONE
PROTECTION DES BORDS EN PROFIL INOX
PROFIL SILICONE
IMPLANTATION DES TUYAUX D'AIR COMPRIMÉ
FIXATION IDENTIQUE AUX VOLUMES DE VITRAGE
voir elevation A
+74.15
+66.05
1988
2000
1738
125
4 x 2.0m (=8.0m)
8.1m
4 x 8.1m (=32.4m)

MAINTENANCE EQUIPMENT

The maintenance of a glazed building as large as the Serres is an important consideration which had to be taken into account at the design phase of the structure.

A suspension rail for maintenance equipment hangs above and below each structural horizontal member of the façade. This rail, a 60 mm diameter tube, is fixed to the structural tubes at every 2 m; it supports the maintenance ladder on small wheels, providing access inside the Serres for cleaning purposes and for replacement of the glass panels.

On the roof, a maintenance platform is fixed to the outside of each Serre. It allows cleaning and repairs as well as the replacement of the roof units. It also supports a cradle that travels outside the façade to allow the cleaning and inspection of the silicone joints. A second maintenance platform, located under the roof, gives access to the entire inner surface: a small cradle hanging from the truss girders can slide the length of the Serre.

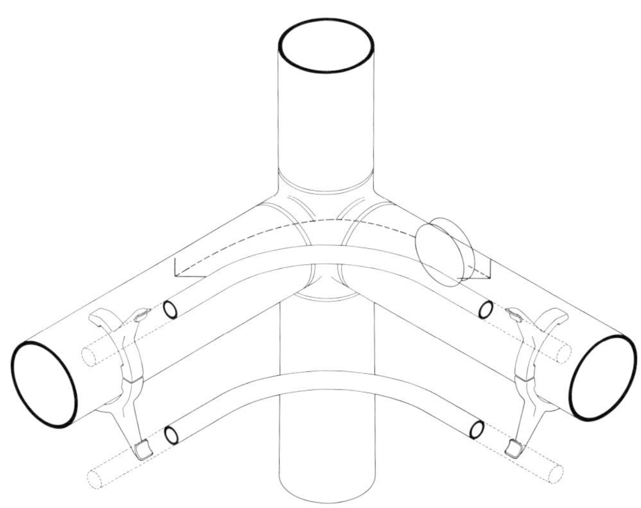

Above:
A Serre's maintenance ladder and the maintenance rails coupled to the horizontal mullions of the framework.
Left:
Attachment of the maintenance rails to the framework.
Right-hand page:
Working drawings of a Serre's maintenance equipment.

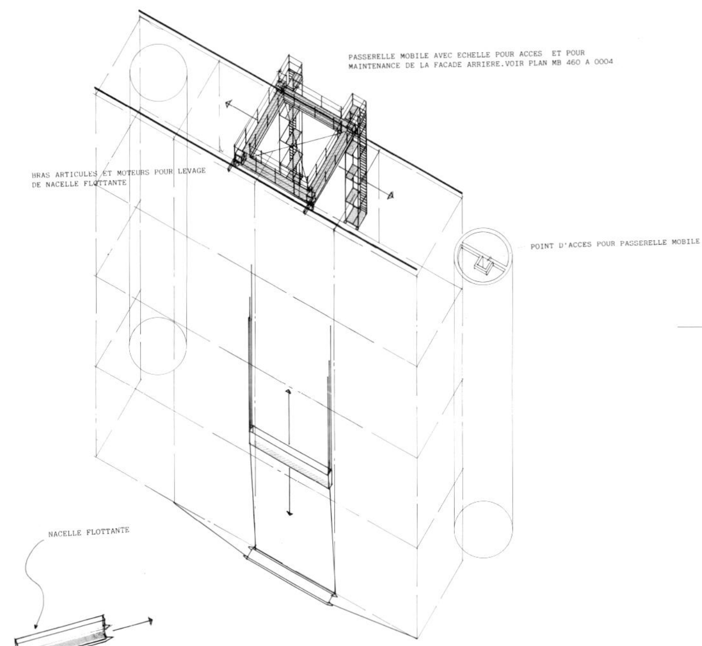

MAINTENANCE DE L'EXTERIEUR

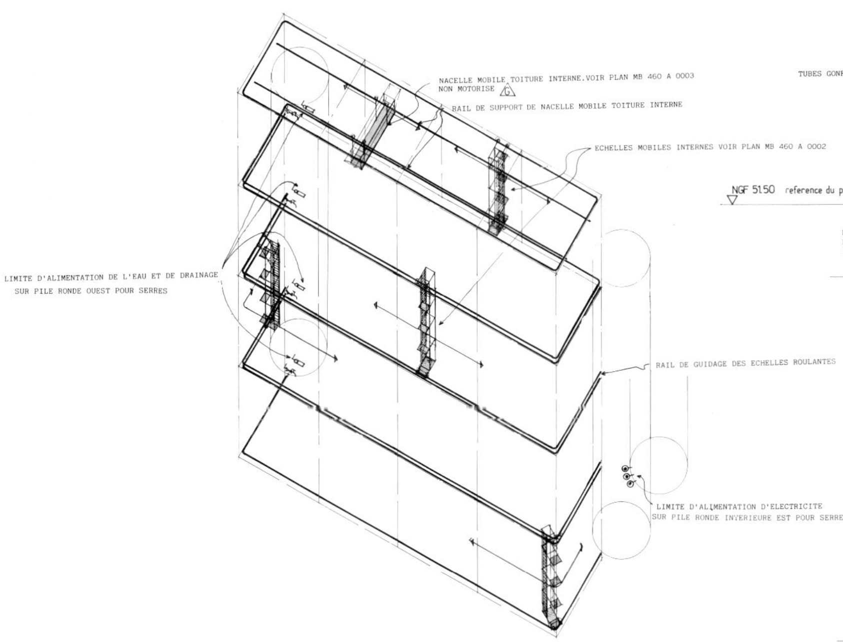

MAINTENANCE DE L'INTERIEUR

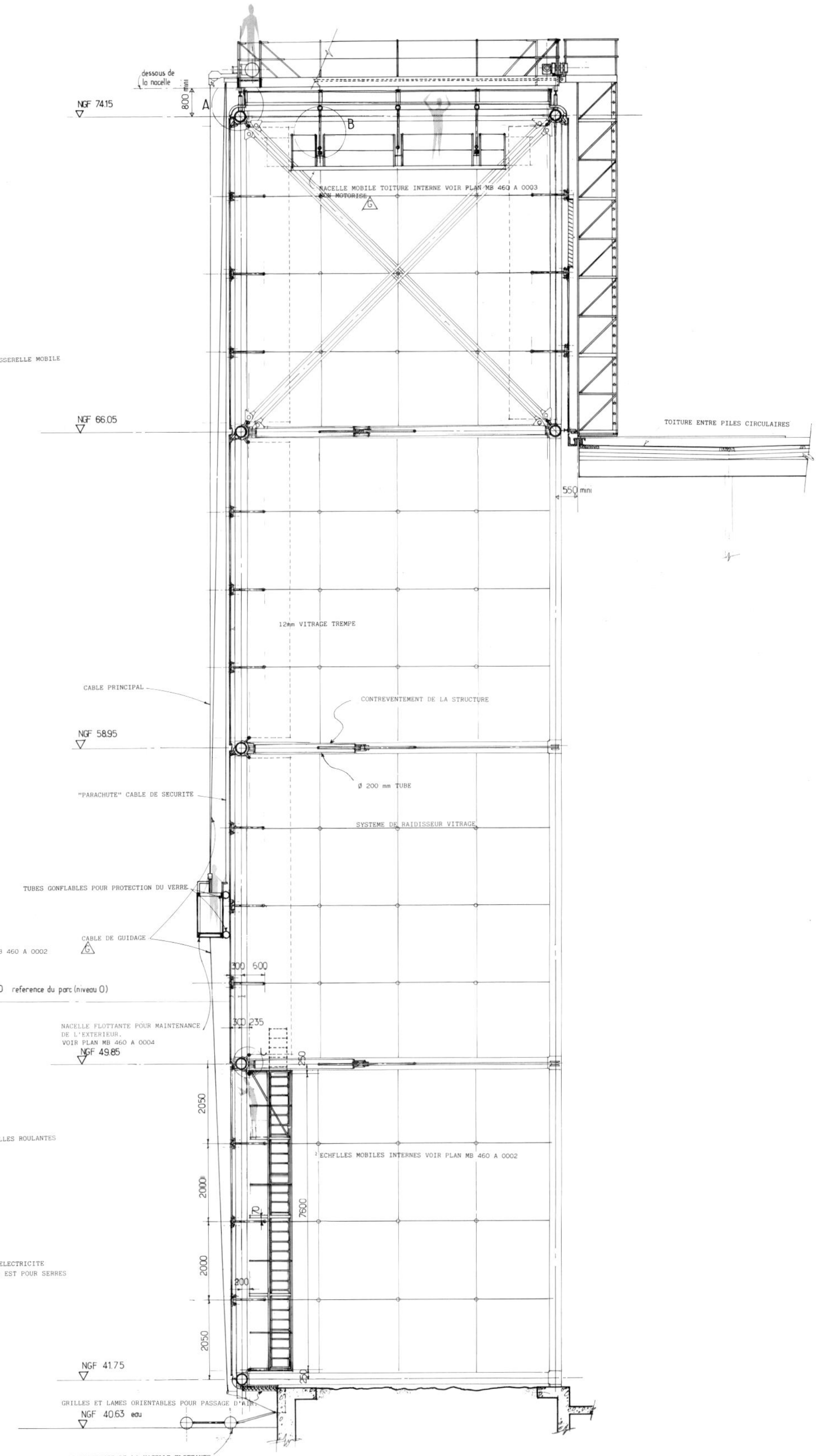

COUPE GENERALE

SPECIFICATIONS

The principle of structural glass requires detailed specifications to be drafted for the glass and the articulated bolt, and these are reproduced below.

Glass

Basic float product

The glass structures are manufactured in float glass free from impurities and discolorations. The light transmission of the glass is 75% and its average reflectivity is 12%. The tolerance in the glass thickness must be less than 0.05 mm.

Cutting and toughening

• Cutting

The use of a numerically controlled cutting machine is recommended for cutting the glass, to achieve precise alignment of the edges. The edges are ground smooth with chamfers on the corners.

• Drilling

The holes are drilled with a diamond drill bit to a coaxiality tolerance between the conical and cylindrical parts of less than ±0.1 mm. The base of the cylindrical holes is chamfered. It is recommended that the holes be positioned in relation to one another, and not in relation to the edges, to a precision of ±0.1 mm using a numerically controlled tool. There must be no flaking or shoulders inside the bore holes. The internal surface of the holes must be smooth, with minimal roughness resulting from the drilling.

• Toughening

The glass is preferably toughened horizontally and heated to 700°C, then carefully cooled to achieve a regular distribution of toughening stresses in order to minimize the surface deformations caused by differential cooling. This level of toughening guarantees a minimum working strength of 50 MPa.

• Heat soak

Each glass sheet is heat soaked to minimize the risk of spontaneous fracture due to the sulphide and nickel content. The heat soak process is carried out in accordance with DN 18516, Part 4, which specifies an 8 h process at between 280 and 300°C.

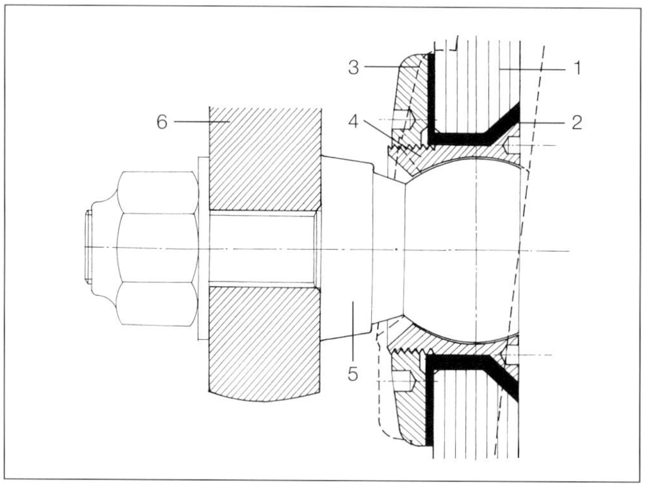

Schematic section of an articulated bolt.
1 Glass
2 Aluminium liner
3 Support washer
4 External collar
5 Stem of the articulated bolt
6 Support

Articulated bolts

Bolts

The bolts are manufactured in stainless steel according to the following standards: stem and ball: F14; ball joint external collar and support washer: 316L.

The hinged connection is assembled by pressing back the external collar onto the ball, rather than by welding or mechanical assembly processes; this avoids shoulders and prevents the possibility of exceeding manufacturing tolerances, which may affect the contact surface. The contact surfaces are lubricated with a PTFE-covered fibreglass cloth. The minimum hardness difference between the external collar and the ball is 50 Shores.

The maximum angle of misalignment of the ball and stem in relation to the axis perpendicular to the glass pane is 5°. A UV-resistant rubber ring, compressible up to 40% of its diameter, is placed between the stem and the collar to ensure self-centring.

Aluminium liner

The insulating aluminium liner, which prevents contact between the glass and the bolt, is annealed to grade 1050. It is mounted with a 1 mm silicone bead to guarantee that the assembly is watertight.

Assembly

The support washer is mounted on to the external collar and tightened to a minimum torque of 100 N m.

CHRONOLOGY OF THE SERRES PROJECT

1980
Architectural competition for the 'Cité des sciences et de l'industrie' won by Adrien Fainsilber. The Serres are an integral part of the project.

1981
The structural concept and the first ideas for the glass suspension system are defined in the preliminary phase of the 'project definition'.

September 1982 to March 1983
Detailed development of the structural concept.

March 1983
The cable truss concept is adopted.

January 1983 to May 1984
First research work on the glass suspension system.

October 1983
Scale model of the cable truss and glass suspension systems exhibited at the Pompidou Centre.

January 1984
Contract negotiations with CFEM.

January 1984 to July 1984 Design development of main structure node castings with industry.

April 1984
The spherical bearing articulated bolt is adopted.

May 1984
Working drawings of the glass suspension system completed.

June 1984
Glass testing begins. Visits to the Bourget Air Show (Salon du Bourget).

July 1984
Contract signed with CFEM. Manufacture of main castings and structural tubes begins.

December 1984
Testing of prototype of a full 8 × 8 m panel.

January 1985
On site work on the Serres begins.

February to March 1985
Development and testing of the aluminium liner.

April 1985
Erection of the tubular structure.

May 1985
The glass suspension system is approved by the building control engineers (Bureau de contrôle).

September 1985 to January 1986
Erection of the glass system.

December 1985 to March 1986
Ventilation and smoke exhaust systems installed.

March 1986
Museum opened by President François Mitterrand.

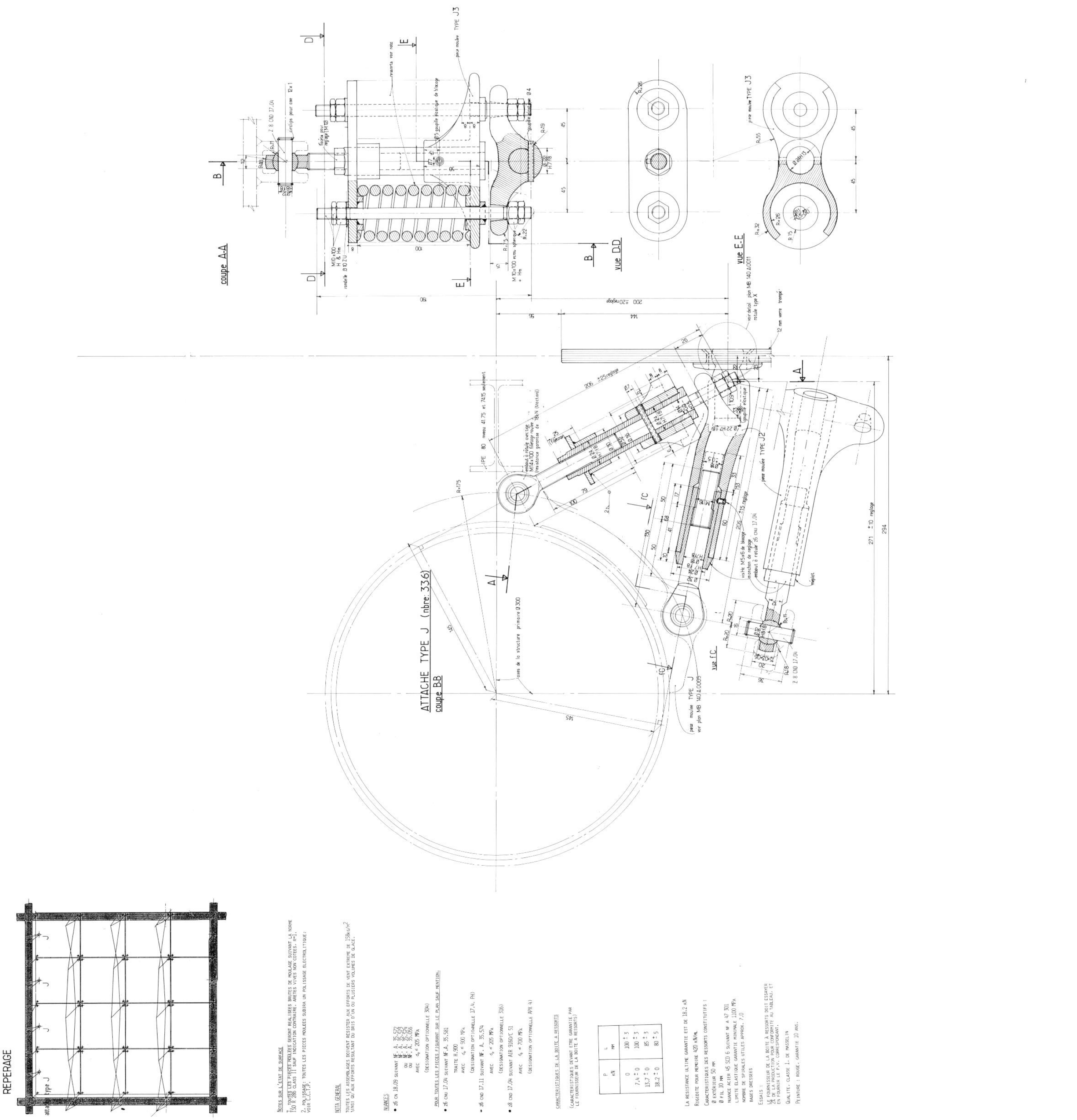
ATTACHE TYPE J (nbre 336)
coupe B.B
coupe A-A
vue D.D
vue E.E
vue f.C
REPERAGE

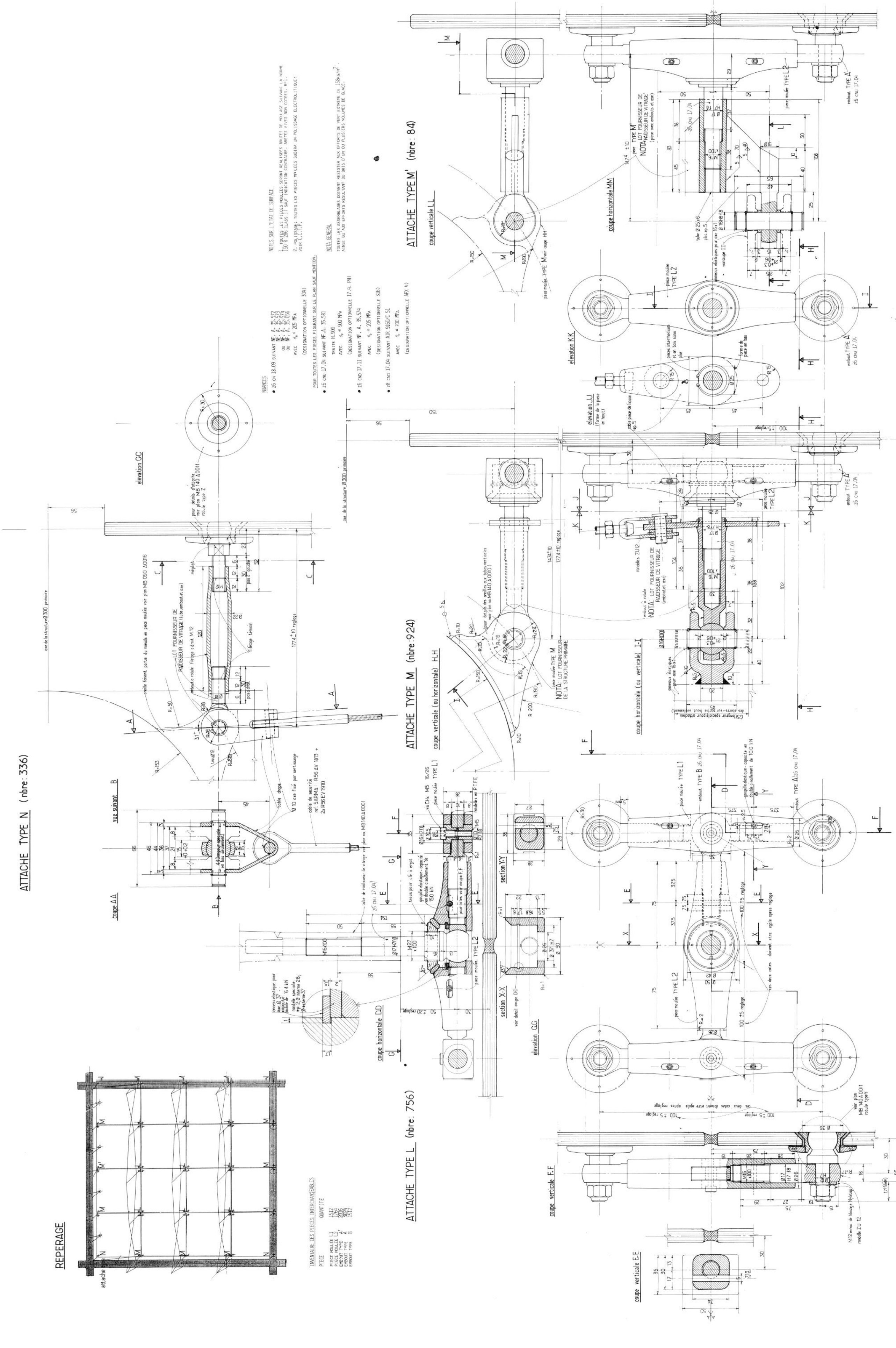
ATTACHE TYPE N (nbre: 336)
ATTACHE TYPE L (nbre: 756)
ATTACHE TYPE M (nbre: 924)
ATTACHE TYPE M' (nbre: 84)
REPERAGE

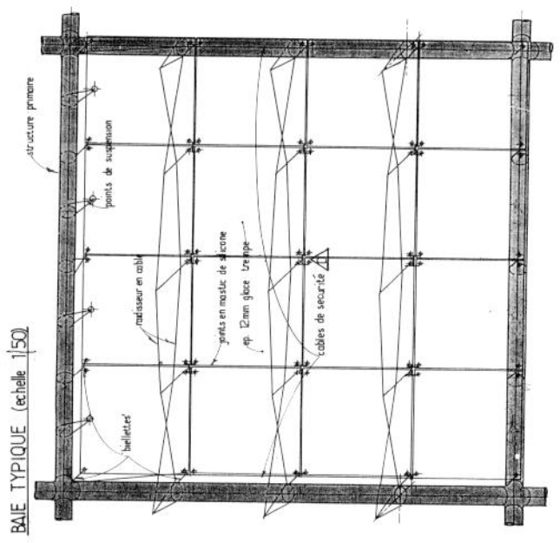
BAIE TYPIQUE (echelle 1/50)

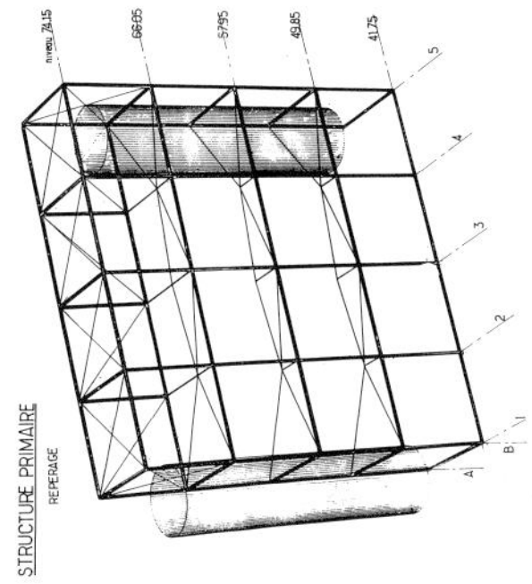
STRUCTURE PRIMAIRE
REPERAGE

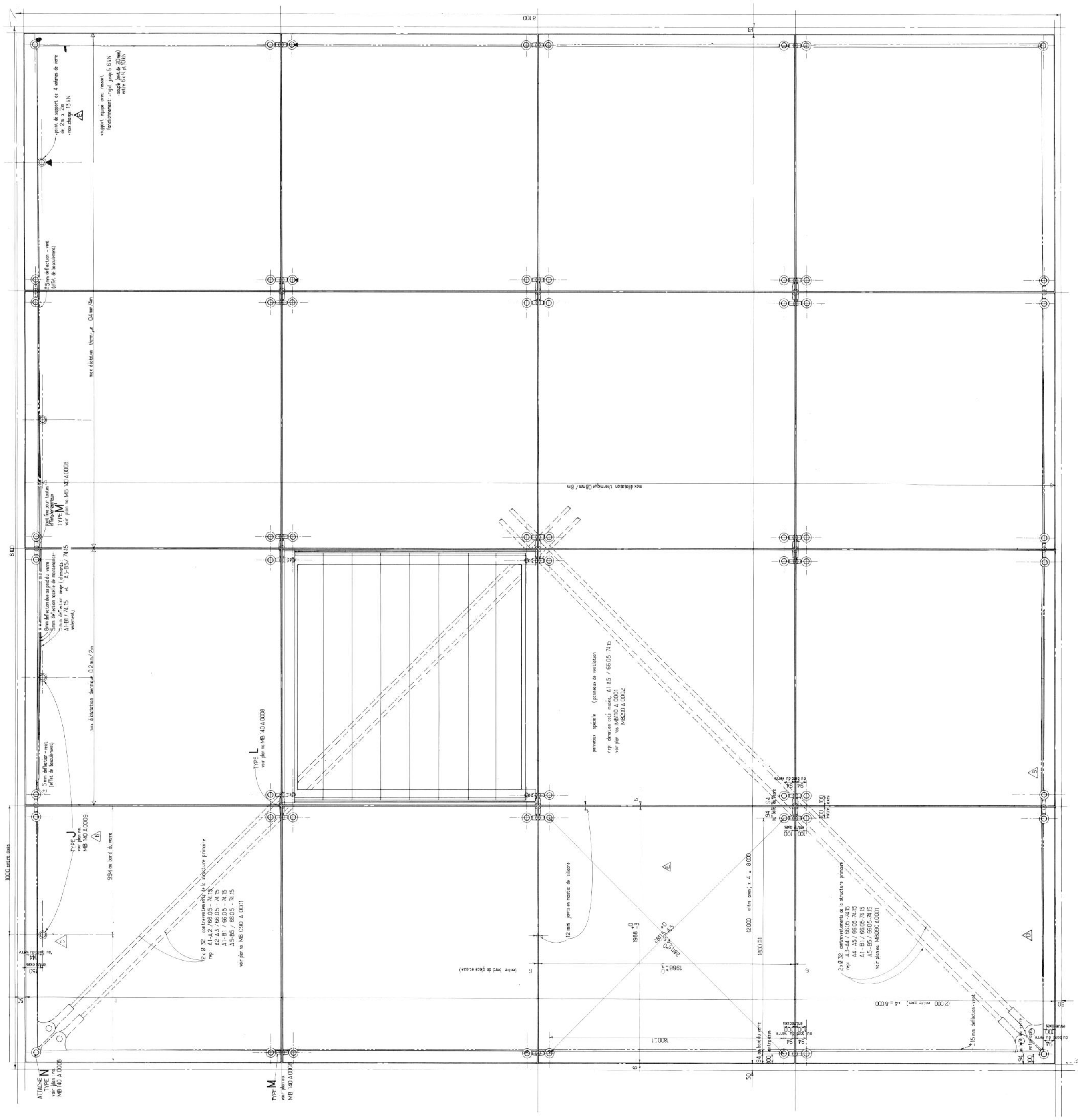

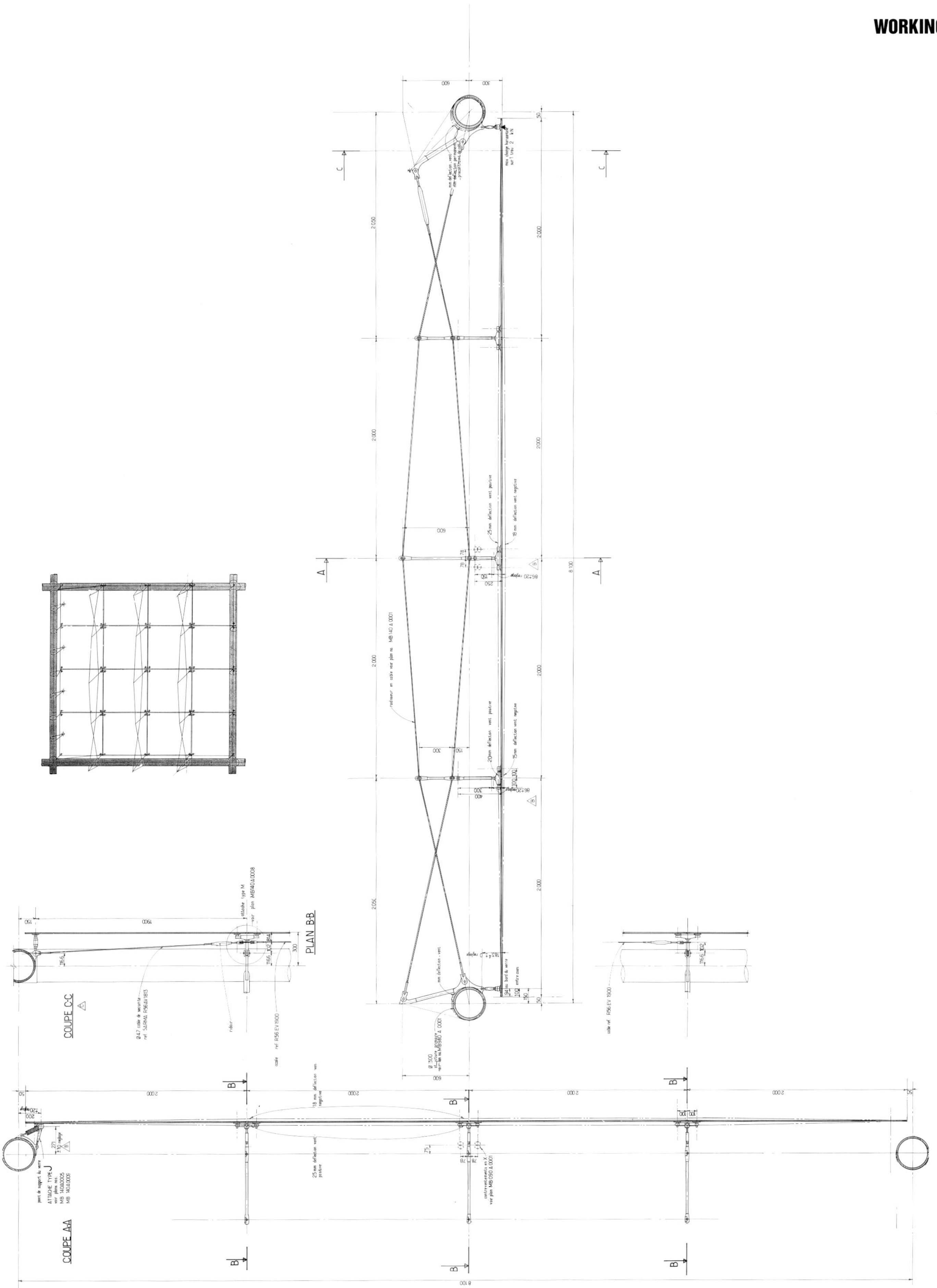
PLAN B-B
COUPE C-C
COUPE A-A

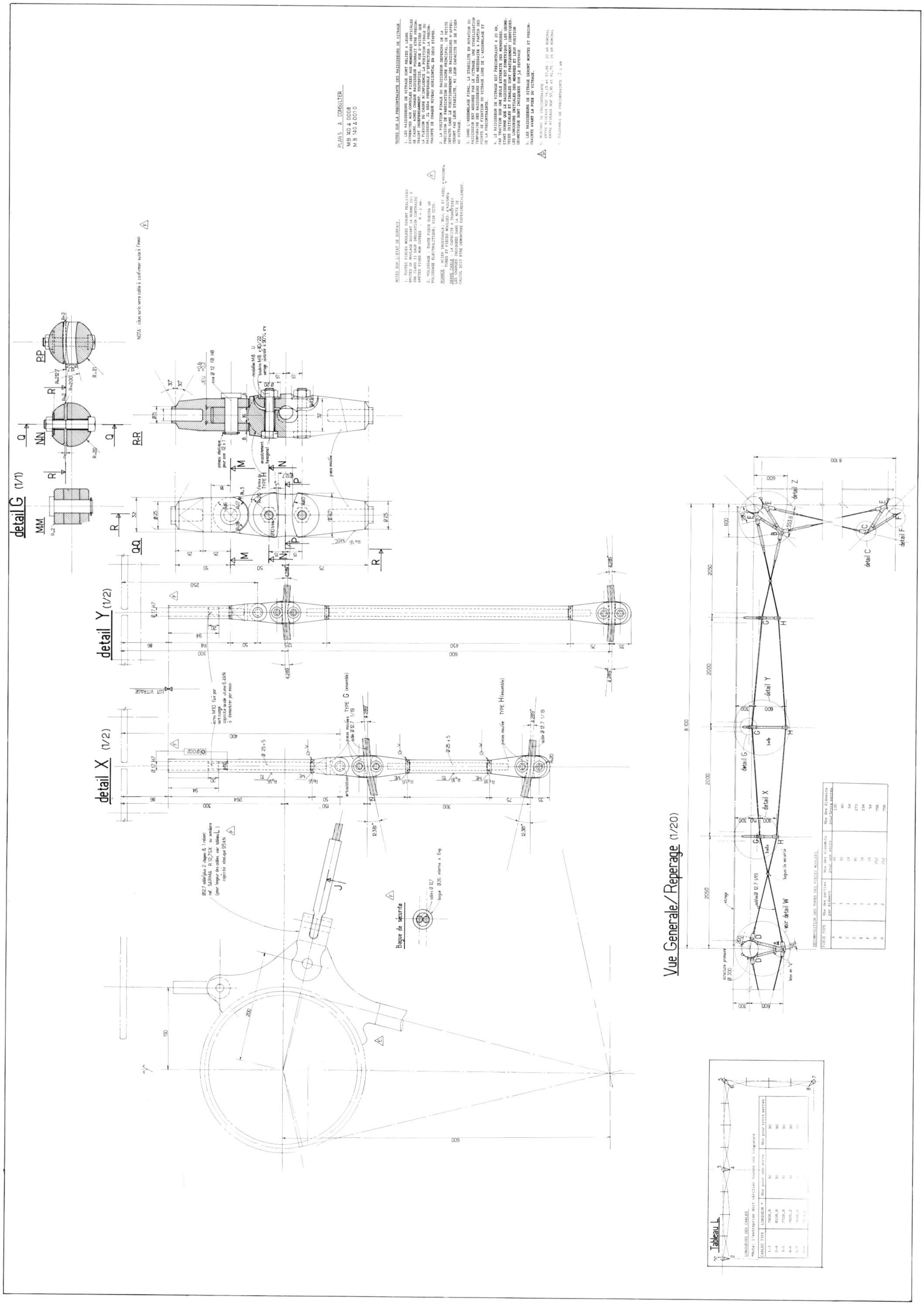
detail G (1/1)
detail Y (1/2)
detail X (1/2)
Vue Generale/Reperage (1/20)
Tableau L
Bague de sécurité
PLANS A CONSULTER

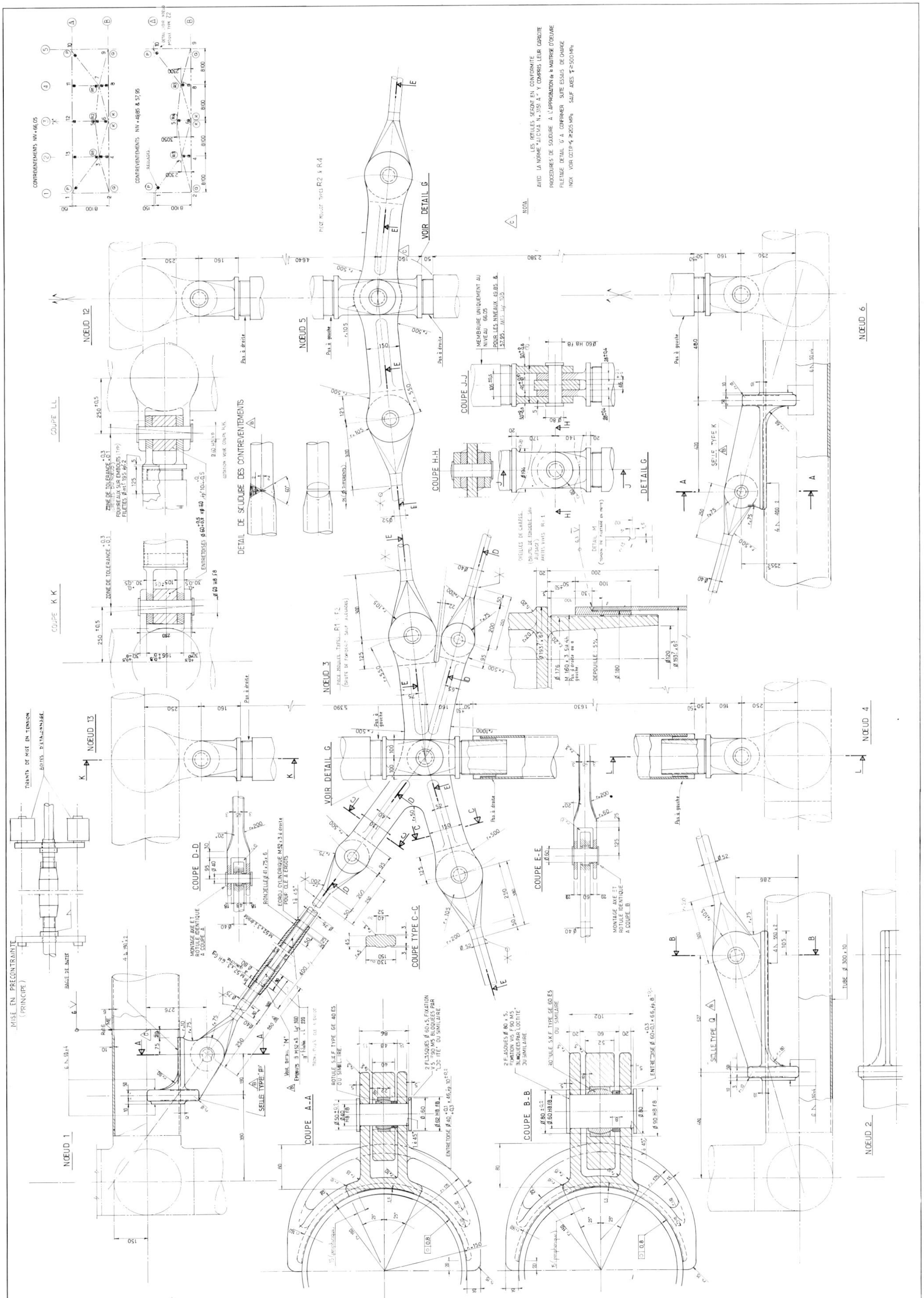

NŒUD 12
NŒUD 13
NŒUD 5
NŒUD 3
NŒUD 6
NŒUD 4
NŒUD 1
NŒUD 2
VOIR DETAIL G
DETAIL G
COUPE H-H
COUPE J-J
COUPE L-L
COUPE K-K
DETAIL DE SOUDURE DES CONTREVENTEMENTS
COUPE D-D
COUPE E-E
COUPE TYPE C-C
COUPE A-A
COUPE B-B
SELLE TYPE K
SELLE TYPE Q
MISE EN PRECONTRAINTE
(PRINCIPE)

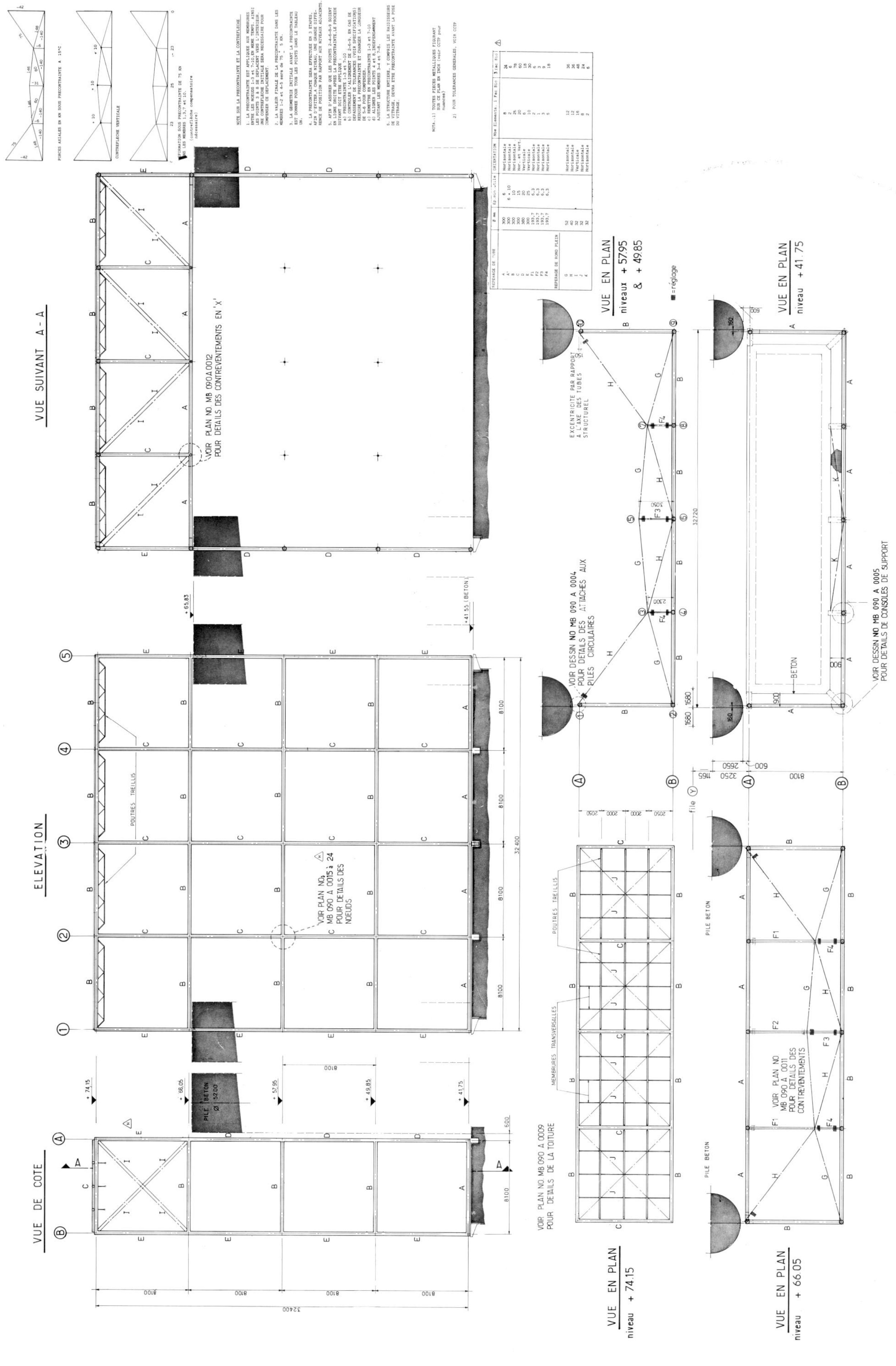

VUE DE COTE
ELEVATION
VUE SUIVANT A - A
VUE EN PLAN niveau + 74.15
VUE EN PLAN niveau + 66.05
VUE EN PLAN niveaux + 57.95 & + 49.85
VUE EN PLAN niveau + 41.75
POUTRES TREILLIS
MEMBRURES TRANSVERSALES
PILE BETON
VOIR PLAN NO MB 090 A 0012 POUR DETAILS DES CONTREVENTEMENTS EN 'X'
VOIR PLAN NO MB 090 A 0009 POUR DETAILS DE LA TOITURE
VOIR PLAN NO MB 090 A 0011 POUR DETAILS DES CONTREVENTEMENTS
VOIR DESSIN NO MB 090 A 0004 POUR DETAILS DES ATTACHES AUX PILES CIRCULAIRES
VOIR DESSIN NO MB 090 A 0005 POUR DETAILS DE CONSOLES DE SUPPORT
EXCENTRICITE PAR RAPPORT A L'AXE DES TUBES STRUCTUREL

STRUCTURAL GLASS IN OTHER PROJECTS

The ideas that we have just described, which are illustrated in the Serres at La Villette, can of course be applied to other projects. Here we present 15 subsequent buildings based on the same structural glass concept and designed by RFR. We shall look only at the use of glass in them, but other publications describe the projects in detail.

The different projects explore the potential of the plane of glass in architecture in a manner that exploits the technique of the removal of the traditional mullions. In these cases it is the plane of glass itself, with or without luminous treatments such as translucency and fritting, and not the graphic silhouette of mullions, that becomes the subject of architectural expression. It is an immaterial membrane that can be transparent or reflective and that also offers the many ambiguous states that exist between these extremes.

Along with the concepts of process, predictability and hierarchy which guided the design, the theme of transparency is the common denominator of these buildings. As we have seen, transparency can be interpreted as the ambiguity that exists between the presence and the disappearance of the glazed pane. This is the duality of transparency: seeing through something also implies that one is aware of the surface being looked through. In the projects shown here, we examine this subtle threshold between making the glass surface disappear and visually expressing the structures and suspension systems holding it in place. To do this we must add the reflective nature of glass, i.e. the subtle range of variations between absolute transparency and absolute mirror reflectivity.

The projects presented in this Postscript are a further analysis of the determinant role of glass bracing structures in relation to transparency. In most cases, the objective is to reduce the size and visual impact of these structures. Thus, at La Villette the aim of the prestressed cable truss system is to minimize the physical presence of the bracing structure. Yet these structures are still purposefully expressed: there is no attempt to make them disappear. They create a link between the glazed pane and the architecture of the building as a whole and remain part of a precise architectural context. The structures often explain by their geometry the way in which the loads applied to the glass are restrained. This geometry is a response to the configuration of the glazed plane, to the perpendicular wind loads and to the requirements of the main building structure to which these loads are transferred.

Making the path of the loads in the glass obvious through an appropriate geometrical solution serves the purpose of explaining to the viewer how the whole works and how the simple problem of holding a plane of glass and of ensuring its wind bracing has been resolved.

The combination of glass skin and support structure should appear to be a symbiosis of these two elements: the glass skin must need the structure, the structure must need the glass skin. Each must be intrinsically interdependent, while maintaining its own integrity.

The 15 projects that follow also observe the same design and fabrication processes as the Serres at La Villette.

First of all, the design of each part of the project is developed and a geometry is chosen that is appropriate for the general architectural concept of the whole building.

Then follows the essential stage of informing those involved in the project. This involves expressing the way it works and the way it will be made, so as to establish the cost as well as to plan both the manufacturing process and the construction work. Regular cooperation with the companies involved is necessary at all stages: in every project there are specific critical points that require more attention than others.

The fundamental principle of predictability applies to the design of every detail of the structure and of the glazing. In most cases, the first step is to know as precisely as possible the loads that can be withstood by each glass unit. All structures move due to temperature changes and deflect in response to loads, but glass is fragile and has virtually no capacity for deformation in its plane and absorption of these movements. If the glazing is bolted directly to a steel structure which moves, it could be subjected to very heavy loads and break. It is therefore important to ensure that the way in which the glass is fixed does not create unpredictable loads in it when the support structure bends and distorts.

The second point – still in the context of the concept of predictability – is the attachment of the wind-bracing systems to the main building. The connections must allow for the play between the movements of the structural glass construction and those of the main building.

Not all the projects discussed in this Postscript make use of the articulated bolt: two of them, the Lintas footbridge and the pavilion at Groningen, use glass structurally but with different fixing techniques, such as patch plate friction assemblies and other more conventional fixing systems.

The projects are presented in chronological order. The Lintas footbridge was built at the same time as La Villette. Then follows the pavilion designed for Groningen by Bernard Tschumi, the atrium or 'Nef' (Nave) of the office building designed by Jean-Pierre Buffi at La Défense and the Banque Populaire de Montgermont by Odile Decq and Benoît Cornette, where a double glazing system was used for the first time. Next we mention the Canopy at Verdun (RFR), the entrance for the CNIT at La Défense and the glasshouses of the Parc Citroën (Patrick Berger). The developments undertaken in Japan in collaboration with Asahi Glass are briefly discussed, particularly with regard to resistance to seismic disturbances and typhoons. The inverted pyramid at the Grand Louvre (I.M. Pei) and various recent projects making the most use of the articulated bolt, such as the TGV terminal at Roissy airport (Paul Andreu), the atria of an office building situated in Avenue Montaigne in Paris and finally Richard Rogers' building for Channel 4 in London, constructed by Eiffel Construction Métallique, are then examined. We end with the project for the west façade of Lille Cathedral, using translucent marble, on which Peter Rice was working before his premature death.

THE LINTAS FOOTBRIDGE, PARIS

Date: 1985. Architect: Marc Held. Client: Lintas Paris. Building control engineers: Socotec. Contractors: Parisot, general building and structural contractors; Miroiterie Générale, glazing contractor; Adès, neon lighting; Boussois, glass suppliers. Cost per square metre of glazing: FF 6640 excluding tax (1985 values).

The first of several projects 'derived' from the principles applied at La Villette is a footbridge with an 18 m span. The roof, sidewalls and flooring, suspended from a prestressed steel structure, are made of glass, thus creating both reflections and sound reverberation. Within the reorganization of the Lintas Agency it was planned to create an expressive circulation route above the flat terrace roof of the existing building. The brief was to give an image of strong sensations, of energy and of tension through dynamic architecture and technological excellence. Thus, the footbridge is simultaneously a symbol and a practical solution to a problem of space management.

Since the terrace roof did not belong to the client, the loads could only be supported at both ends of the footbridge. The structure therefore comprises two steel tetrahedra supported at the furthest possible anchoring points of the existing building and resting against each other at their meeting point towards the centre of the terrace. The horizontal component of these tetrahedra is a truss girder, which takes up the lateral loads. The floor is suspended from this first assembly. As with any other structure, the prestressed steel structure moves and bends according to the loads it takes up, temperature changes, etc. The glazing must not restrict this movement.

The roof rests directly on the horizontal steel girder. Extruded neoprene joints are fitted onto small V-channels welded to the structure. The glass panes rest on these joints and are held at each corner by clips that prevent them from lifting.

The sidewalls are suspended from the same horizontal girder by articulated fasteners. Each pane is suspended by just one triangular fastener and one vertical one. This assembly holds the glazing in place but does not restrict deformation of the structure.

The footbridge spans an 18 m void. The effects of the reflections in the four glass panes are infinitely multiplied. Below left, fixing details of the glass walls to the steel beam.

POSTSCRIPT

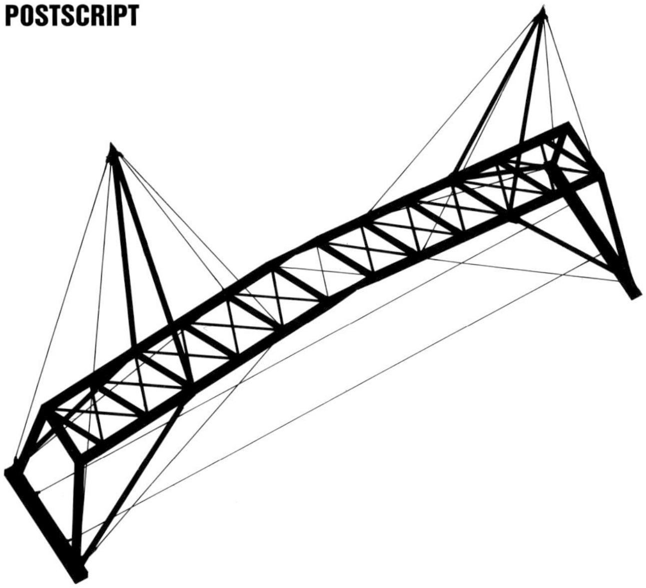

As hinge movement between the two structural tetrahedra in the centre of the footbridge can occur, the central pane of glass is not aligned on the plane of the others; instead it is suspended by two simple horizontal battens. The two halves of the footbridge can thus freely deflect without applying any unpredictable loads to the glass.

The floor is made of laminated glass that has been sandblasted in bands to give it a non-slip surface. Sheets of 500 × 700 mm glass are fitted into a steel angle section frame resting on a flexible neoprene support. This arrangement allows a good load distribution around the periphery of each panel, thus avoiding a concentration of stress.

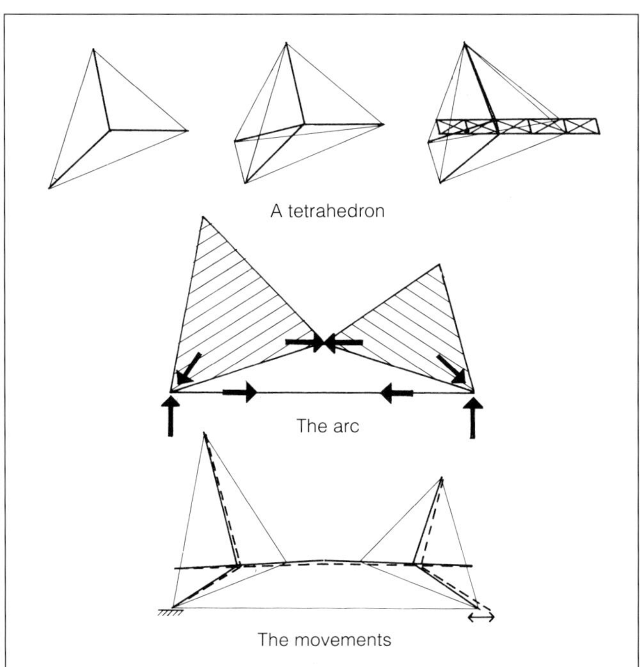

Left:
Diagrams showing the principles of how the structure works.
Above left: Axonometric of the tetrahedra and the structural truss from which all the other elements of the footbridge are suspended.

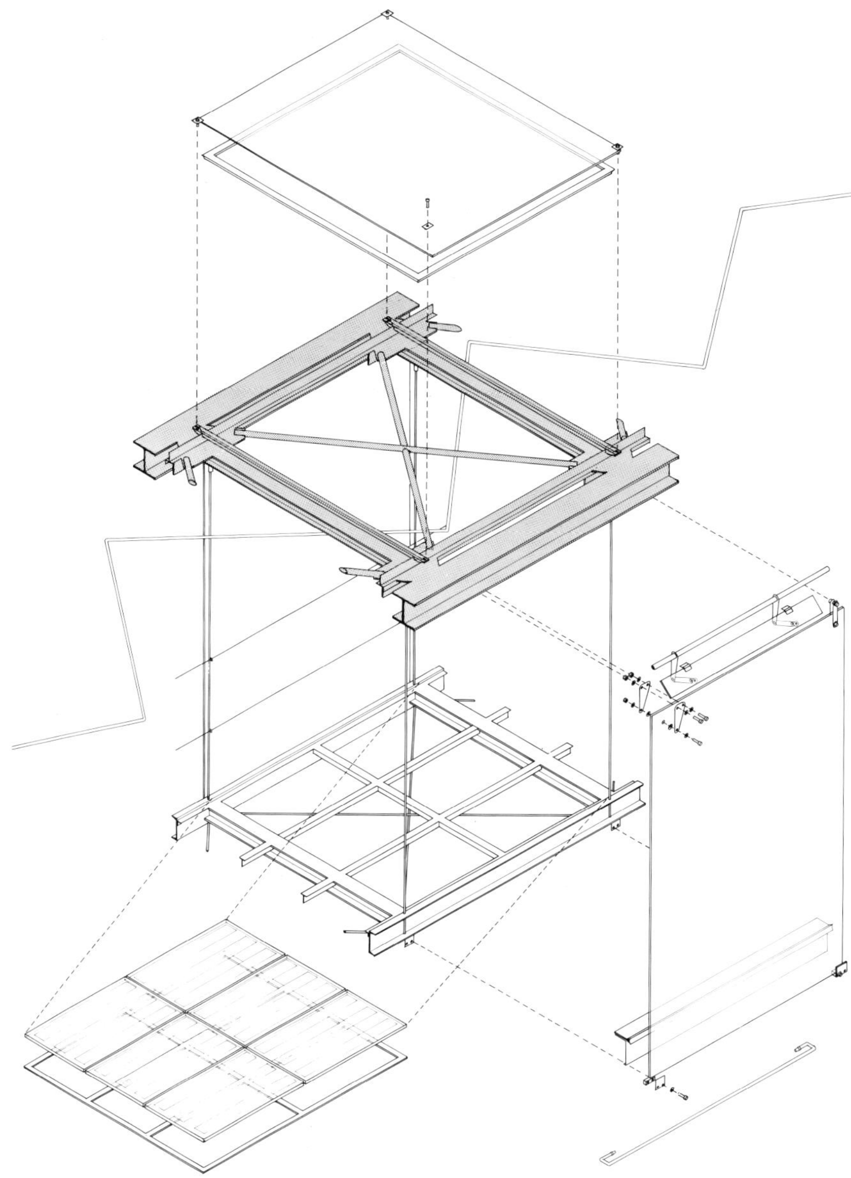

Exploded axonometric of a typical bay.

THE GRONINGEN VIDEO PAVILION

Date: 1988. Architect: Bernard Tschumi. Client: the town of Groningen, The Netherlands. Estimated cost per square metre of glazing: FF 5000.

This small video pavilion, situated in one of Groningen's public parks, is an example of structural glass used largely in bending, rather than in suspension as at La Villette and in the other examples discussed here. In the pavilion, glass fins are used for the very reason that they are rejected at La Villette: for their reflectivity, which creates a visual barrier that partially obstructs the transparency.

In this case, the multitude of reflections of the ambient light mixes with the direct and reflected light from the video screens and is exploited for its visual interest.

The glass technology is simply the application of conventional patch plate friction assemblies in traditional glass fin glazing.

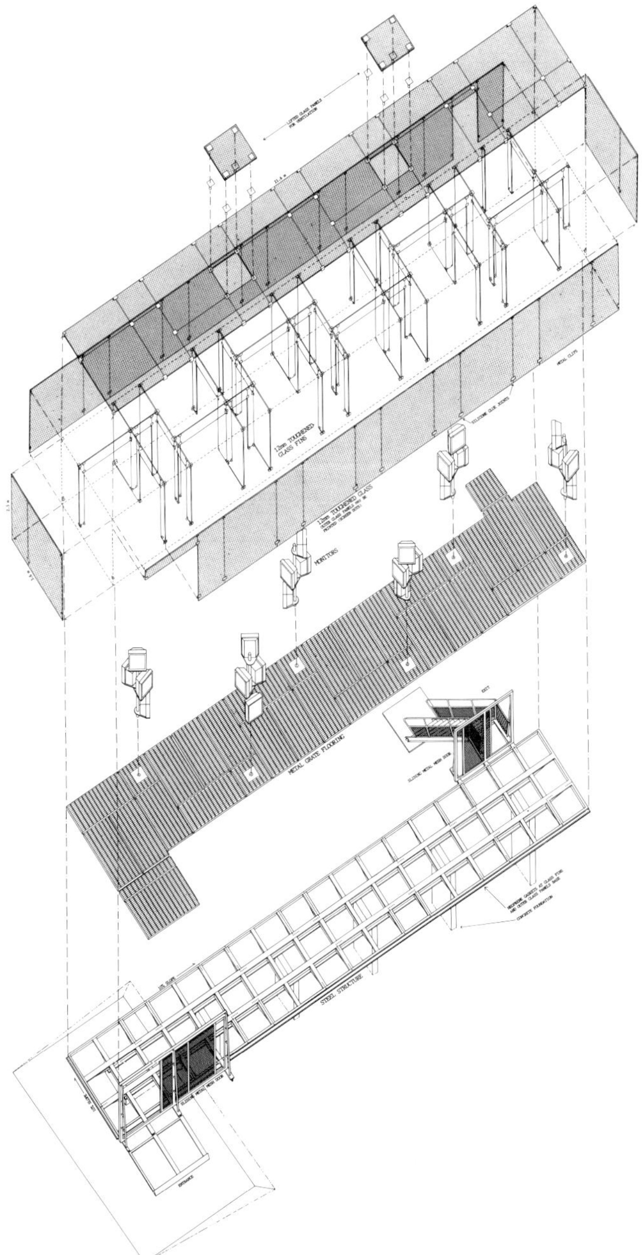

Left:
The 21.6 x 3.6 x 2.6 m parallelepiped, entirely of glass, is inclined by 10° from the vertical and horizontal. This tilting, combined with the multiple reflections, creates a troubling off-balance sensation.
Above:
Exploded axonometric of the various parts of the pavilion.

THE 'GRANDE NEF' OF THE COLLINES DU NORD – LA DÉFENSE, PARIS

Date: 1990. Architect: Jean-Pierre Buffi. Client: Sem Tête-Défense. Building control engineers: CEP. Contractors: Viry, structure; Dutemple, glazing contractor; Solemat, maintenance systems; Saint-Gobain, glass suppliers. Cost per square metre of glazing (including structure, glass and maintenance): FF 5950 excluding tax (1988 values).

The Florentine architect Jean-Pierre Buffi, designed the office buildings called 'Les Collines' as a series of rectilinear blocks in rows on each side of the Grande Arche at La Défense. The complex on the north side of the arch is interconnected by a single glazed internal street, the Nef (or Nave), which slices through the rows of offices at a slight angle. Inside the Nef, the lift towers are left exposed.

From a structural point of view, the Nef comprises a monolithic roof resting on the rows of office blocks, from which the glazed lateral walls corresponding to the void between buildings are suspended. The roof, 16 m wide and 90 m long, rests on two lateral longitudinal beams linked every 2 m by fish-shaped trusses. The side walls are wind braced by 16 m long, horizontal Vierendeel trusses (corresponding to the distance between two rows of office blocks), each suspended from the one above it at levels corresponding to the offices' floor slabs (every 3.80 m) and fitted outside the Nef. The general aim is to create an orthogonal grid allowing direct suspension of the glazing without secondary framework.

The roof comprises two glazed surfaces. The upper surface is of transparent glass panes fitted directly onto the fish-shaped trusses on aluminium extrusions. The lower surface, of laminated glass with a polybutyral (PVB) opaline liner, is suspended from the structure and makes a translucent screen that plays a dual role: solar protection by day, artificial light reflector by night. In sunny weather this luminous veil becomes a sort of Chinese lantern onto which the shadows of the trusses and of their fan-shaped ties are projected.

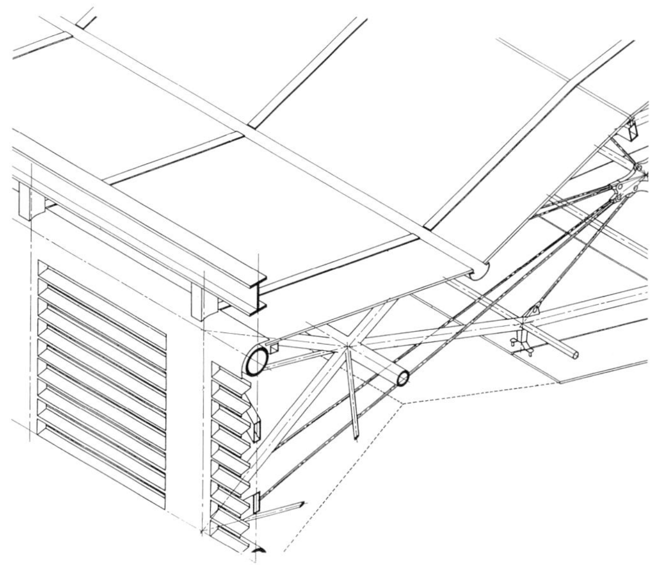

The glass panes of the side walls are suspended directly from the 1 m wide Vierendeel trusses using cast nodes incorporating small suspension 'fingers' fitted with articulated bolts similar to those designed for the Serres at La Villette. Despite their fairly poor structural performance (the absence of a diagonal member means that their shear strength is ensured only by the rigidity of the nodes), the Vierendeel trusses were retained for aesthetic reasons – the architectural project includes no triangular shapes – and also for maintenance purposes – the glazing cleaning cradles travel inside them.

Right:
Top, axonometric of the lateral end of the roof.
Bottom, diagram showing the principles for maintenance: a, of the roof and exterior façades; b, of the space between the roof and the screen; c, of the interior façades.
Below:
Axonometric of the office buildings (the north and south Collines) conceived by Jean-Pierre Buffi on either side of the Grande Arche. The Grande Nef runs through the north Collines.

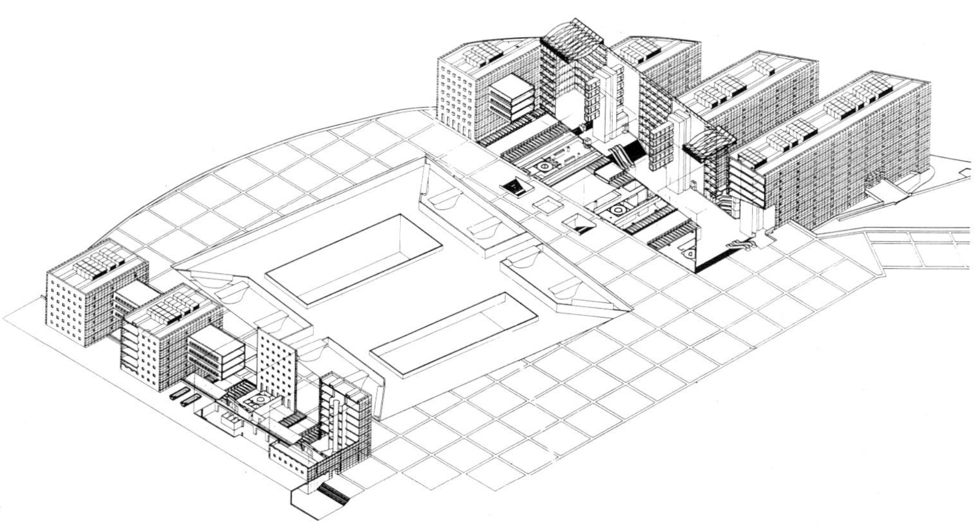

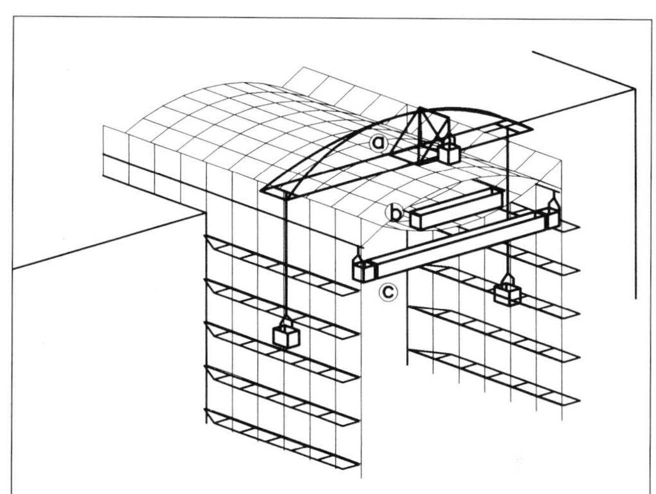

The Nef creates an axis crossing the buildings. The roof screen is brought to life by the interplay of shadows cast by the structure, while the façades almost disappear because of the horizontal positioning of the radial beams, which minimize their visual presence.

BANQUE POPULAIRE DE L'OUEST ET DE L'ARMORIQUE, MONTGERMONT

Date: 1990. Architects: Odile Decq and Benoît Cornette. Client: Banque Populaire de l'Ouest et de l'Armorique. Building control engineers: Socotec. Contractors: Eiffel, metal structures, general contractors, metal frame and suspension parts; PMB, glazing contractors; Maccoco, glazing suppliers. Cost per square metre of glazing (including structure, glass and solar protection system): FF 7260 excluding tax (1989 values).

This project, undertaken by the architects Odile Decq and Benoît Cornette, furthers the research conducted into suspended glazing and the idea of transparency, while at the same time exploring the field of double glazing. From the architectural point of view, the south-facing façade demanded maximal transparency, but it also required the use of double glazing and solar shielding against direct sunlight.

The idea was to find a solution that expressed an appropriate high technological performance in the given natural conditions. In this particular case, the façade had to provide solar protection, while remaining as transparent as possible. In fact, it responds to the variable height of the sun in the sky as well as to the frequency of exposure.

The structure of the façade both supports the brise-soleil and wind braces the glazing system. It is placed outside the building itself, well in front of the glass curtain, which is suspended from steel trusses in the building's parapet wall. The façade is divided into sections 12 m long and 8 m high.

The glass curtain structure has three components: tubular masts placed every 12 m; horizontal *sablières* trusses placed on top of the masts; and three standard horizontal trusses. The wind bracing is achieved by using both standard and *sablières* trusses. These three components have the same configuration: a central tube working in compression with associated ties and spacers. When a wind load is applied to the trusses, the assembly reacts as follows. The ties situated on one side of the truss are in tension while the central tube is compressed; when the wind changes direction, the ties on the opposite side of the truss come into play. The shear capacity is achieved through the use of small diagonal ties made of non-prestressed solid rods. Rigging cables in an 'inverted V' arrangement hold the trusses in their horizontal plane and thus stabilize them against buckling and

The main façade of the bank, which mixes metal and glass, clearly affirms its high technical excellence along its 120 m length.

overturning. Small articulated struts, 25 mm in diameter and 1.5 m long, link the glass plane to this structure.

Two types of glazing are used: single glazing on the west half of the façade (the entrance hall, which does not require insulation) and double glazing on the east side (offices). The design of the single glazing is identical to that of the Serres at La Villette, except for the two- or four-point suspension pieces made of monobloc cast parts without adjustable eye bolts. They are either star-shaped (four point suspension) or bar-shaped (two point suspension). These bolted assemblies are adjusted at either end by indented washers. The hole of the female component is overdimensioned so that the articulated bolt stem can be freely positioned; the bolts are held in their final position because the cast parts and the washers are indented and these grooves are imbedded into each other when the assembly is tightened in the required position. The assembly is thus simpler than that at La Villette: some of the fairly complicated mechanical components are no longer required and the delicate machining of the cast component threads is therefore avoided. This assembly is used in a similar fashion for both the single and the double glazing. The use of suspended double glazing – something that, to our knowledge, had never been tried before – is the part of the project that represents a particularly interesting development in the field of glazing technology. The holes themselves are the most sensitive part of the suspended double-glazing system.

We have already seen that it is necessary to toughen the glass in order to compensate for its fragility, particularly around the suspension holes, and that the holes must be drilled before the toughening process. When assembling the glass sheets that make up a double-glazed unit, the four holes in a sheet of glass must be perfectly aligned with the four holes of the opposite sheet. There is no margin for error in the precision required in the drilling process. Furthermore, the toughening must be carried out with the glass horizontal, to prevent the holes from moving from their initial positions. Some of the glazed units used in the Bank building are manufactured in laminated glass, made up of three sheets of toughened glass each with four holes - this required the perfect alignment of 12 holes.

The company that manufactured these glass units persuaded us that such a critical assembly was possible without creating the shoulders that result from errors in drilling precision. The holes are positioned using a numerically controlled glass-cutting tool with

Below:
Ground floor plan and transverse section of the entrance.

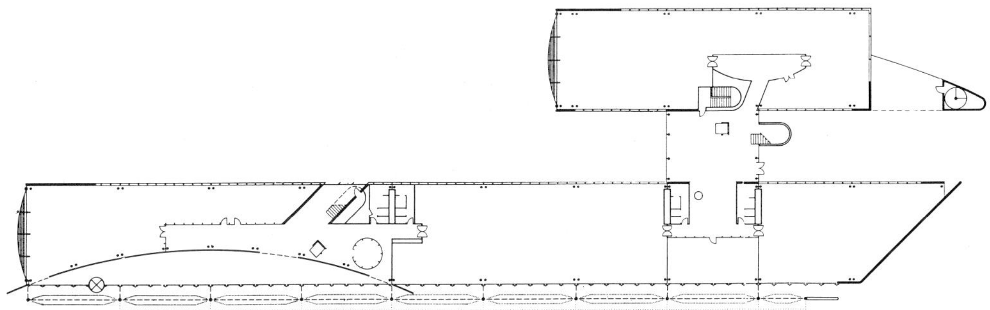

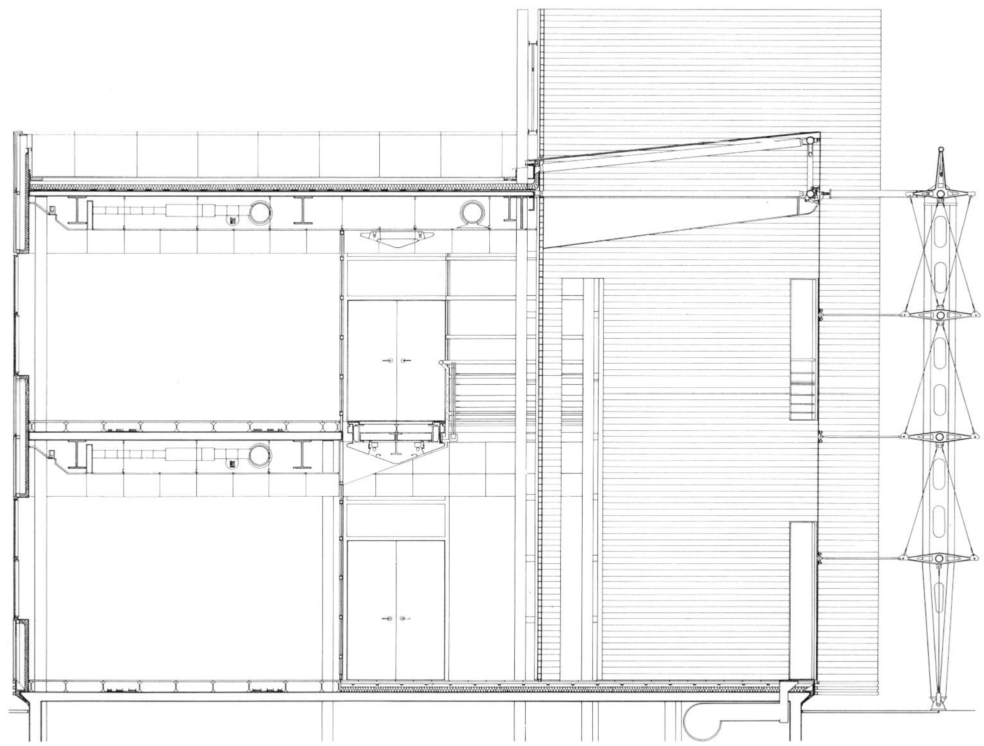

As the structure of the glazed façade is on the exterior of the building, the bare interior is perfectly smooth and the relation between the outside and the inside is marvellously ambiguous.

a negligible error in positioning. The computer controls the positioning of a small tapping hole at each corner of the glass sheet. The toughened sheets are then assembled on to four aluminium rods, which represent the four bolts. If there is any difficulty in positioning the unit on all four rods at the same time, it is rejected.

The seals around the holes between the two sheets of the double-glazing unit include aluminium spacers and sealing mastic in silicone and butyl sealing compound. The spacers are hollow and filled with small pellets of absorbent material. This aluminium spacer and sealing compound system was adapted for the project by the manufacturers of the glazing units.

The articulated bolt follows the principle used for the single glazing, but its dimensions are different: its articulated head is thicker so that it can support the

two sheets of glass. Tension tests have confirmed that the structural performance of the double-glazing assemblies is comparable to that of the single glazing.

The Bank's façade is particularly interesting as an exercise in transparency. The structure of a façade is normally positioned inside the building or at least it is incorporated in the glass plane. Here the structure is on the outside, 2 m in front of the glazed plane. This inversion gives the impression that the inner space is extended towards the outside and towards the structure itself: the glass appears to have dematerialized.

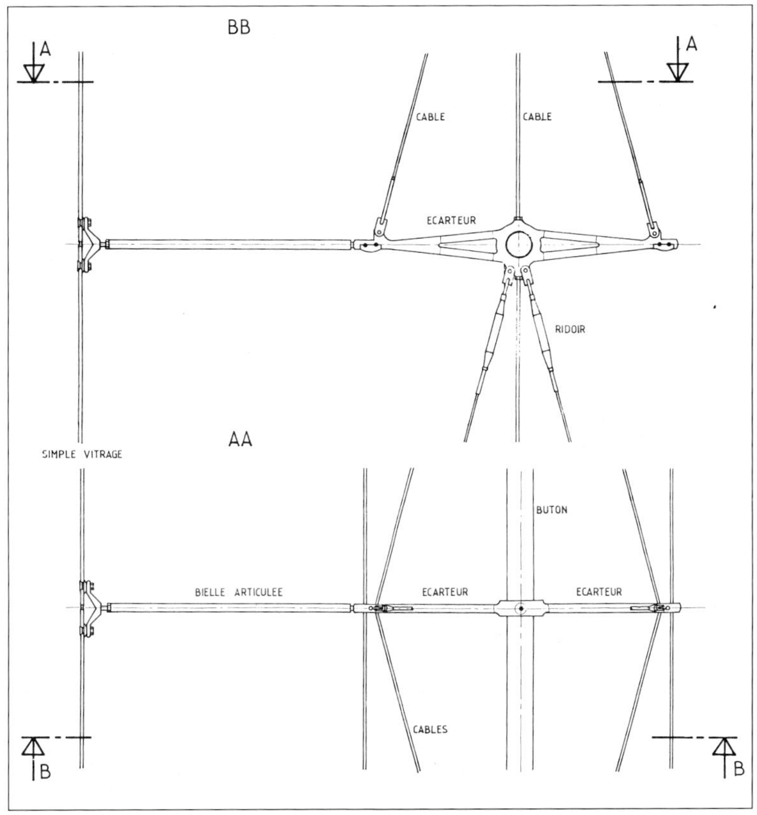

Section and plan of detail on the glass façade and, on the exterior, of the support system for the brises-soleil (not shown) and the articulated struts linked to the glass suspension components.

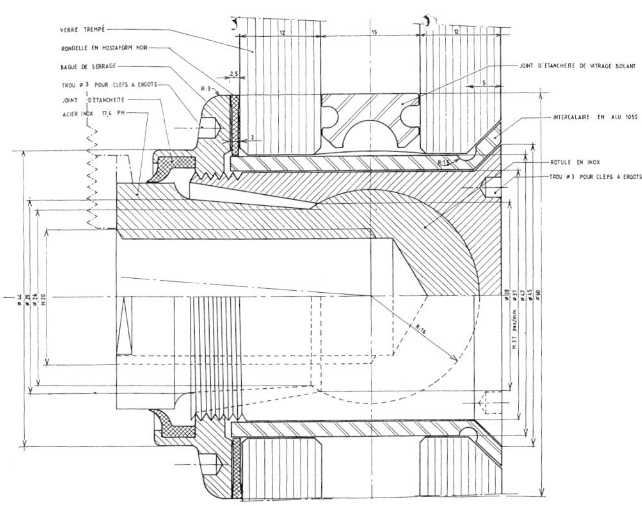

The fixing system developed for La Villette was subjected to tests following its adaptation for double glazing.
Top:
Working drawing of the articulated bolt.

Exploded axonometric of the star-shaped monobloc fixing with the articulated bolt (for double glazing).

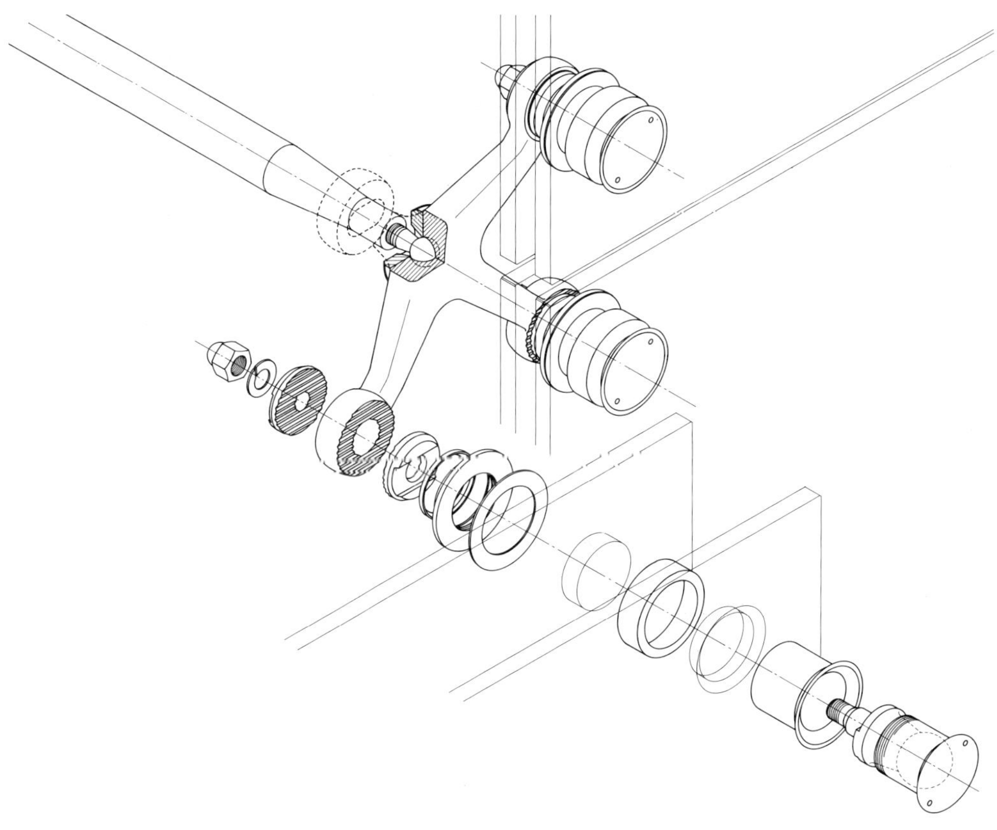

PUBLIC CANOPY, VERDUN

Date: 1990. Architects: RFR-Peter Rice with Roustang, Cuzon and Grandgeorge. Client: The town of Verdun. Consultant to the client: Pierre Colboc. Building control engineers: Véritas. Contractors: Banneel, general and structural contractors; Maccoco, glazing manufacturers. Cost per square metre of glazing (structure included): FF 6600 excluding tax (1988 values).

The focal point of a small open space on the bank of the Meuse is a steel and glass canopy, 160 m long. It symbolizes a brighter future, leaving to the past the sinister memories of World War I. The typical section of the canopy has two parts: a fairly narrow flat part on the street side and a gently curved part on the river side. This formal duality is also structural: a ladder-frame truss is juxtaposed with curved tubular arcs, suspended from ties. The assembly is attached to slightly inclined masts erected every 5 m. The canopy widens in the centre to form a small band stand.

The underside of the canopy is clad with silk-screened, fritted glass sheets measuring 1.5 × 1.3 m, directly attached to the structure by articulated bolts similar to those designed for the Serres at La Villette. To prevent unpredictable stress in the glass due to deformation in the structure caused by the wind, the suspension devices are spring loaded, ensuring that all four suspension points remain in the same plane. The silk-screening of the glass enhances its dual nature of transparency and reflectivity. By day the canopy cover seems transparent yet haloed with a kind of white mist. By night white spots diffuse the artificial light; their brightness increases the visual presence of the glass plane, which itself becomes a light source.

Half way along the canopy the roof widens to shelter a small kiosk.

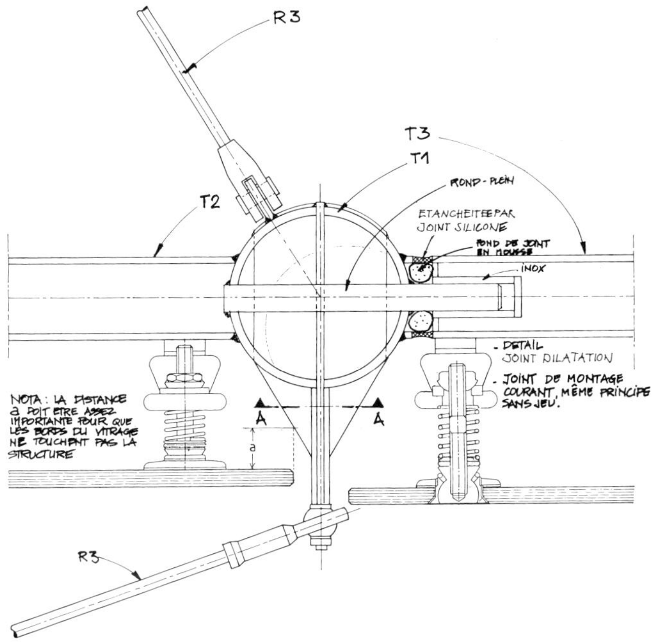

Working sketch and detail of the fixing of the fritted glass, which enables movements of the canopy to be absorbed.

The masts, standing at 5 m intervals, are inclined and articulated at their foot. One mast in three is stabilized by an inclined strut.

NEW ENTRANCE TO THE CNIT – LA DÉFENSE, PARIS

Date: 1991. Architects: Sari Ingénierie. Client: SNC Lucia. Building control engineers: Socotec. Contractors: Banneel. Cost per square metre of glazing: FF 8250 excluding tax (1991 values).

In 1958, the exhibition centre for the Centre National des Industries et des Techniques (CNIT) was completed at the end of a zone of office buildings in La Défense. The centre was designed by the architects Bernard Zehrfuss, Robert Camelot and Jean de Mailly, with Nicolas Esquillan as engineer; the façades were created by Jean Prouvé. In 1988, the architects Michel Andrault and Pierre Parat undertook its redevelopment into a multi-use centre including shops, a hotel and conference halls. The façade on the elevation facing the central *parvis* was protected by its listing as a historical monument, so it was rebuilt exactly as originally

At the foot of the façade of CNIT, on the parvis of La Défense, two covered terraces frame the new entrance to the building.

The glass façade is stabilized by pairs of vertical cable trusses standing on either side of the sheet of glass.

designed.

However, after a few years it became clear that the façade was inappropriate for the new development. The level of the *parvis* had changed and there was no spatial communication between it and the central internal courtyard of the new development. This was because the new façade came straight down onto the *parvis*, at its new level. Prouvé's original design had included a series of doors with an upper level cantilevering overhead to express the entry, but the rebuilt design had no architectural expression of entry except for a series of rotating doors.

The developer, Sari Ingénierie, in agreement with the public authorities responsible for the conservation of France's historical monuments, asked Peter Rice and the RFR team to redesign the façade to maximize its transparency and therefore the spatial communication between the *parvis* and the central courtyard. A further objective was the creation of two small restaurant and café terraces to animate the *parvis* in front of the CNIT.

The terrace cafés are sheltered by glass roofs cantilevered off a curved torsion tube, itself cantilevered through the façade from the structures inside the building. The new entrance is a curved wall set back from a simple hole cut in the façade. By setting it inside, reflections in the glass are minimized, thereby maximizing the views between inside and outside.

Transparency is also maximized by using extra-clear glass (purified of the ferrous oxide that gives it a green tint), as in the pyramids at the Grand Louvre. The glass is hinged from a steel beam fitted in the ceiling space and braced against wind loads with vertical cable trusses, also fixed to the beam. The cable trusses traverse the glass plane, passing through the joints between the glass sheets. This diminishes the visual presence of the trusses, because only half of them are visible from each side of the glass. The glass sheets are suspended from each other by stainless steel H-shaped assemblies and articulated bolts similar to those developed at La Villette, but adapted to permit the cable truss attachment.

The horizontal cross-bar of the H-assembly has been adapted in two fundamental ways: firstly, it includes a central pin, which creates a non-rigid link between the glass and the cable truss that allows independence under differential vertical movements; and secondly, the eyebolt is not threaded, so it is able to slide. This sliding capacity is necessary to reduce the rigidity of the linkages between the vertically suspended glass bands, so that in the event of high wind load from the *parvis*, the curved geometry does not cause the glass to act as a continuous chain, thereby creating extra horizontal loads in the connection. The small amount of expansion possible allows us to predict that the cable truss and not the glass chain will resist the load.

The door itself had to be as minimal as possible, so as not to disrupt the transparency of the façade. Thin, complete doorframes, mounted on articulated bearings, are connected to the cable trusses by the same hinged connections as used on the suspended glass. Thus, in the design system they could be considered as sheets of glass (although resting on the ground rather than being suspended). They do not resist wind load independently of the cable trusses. The door hinging and spring hardware are discretely contained within the doorframe.

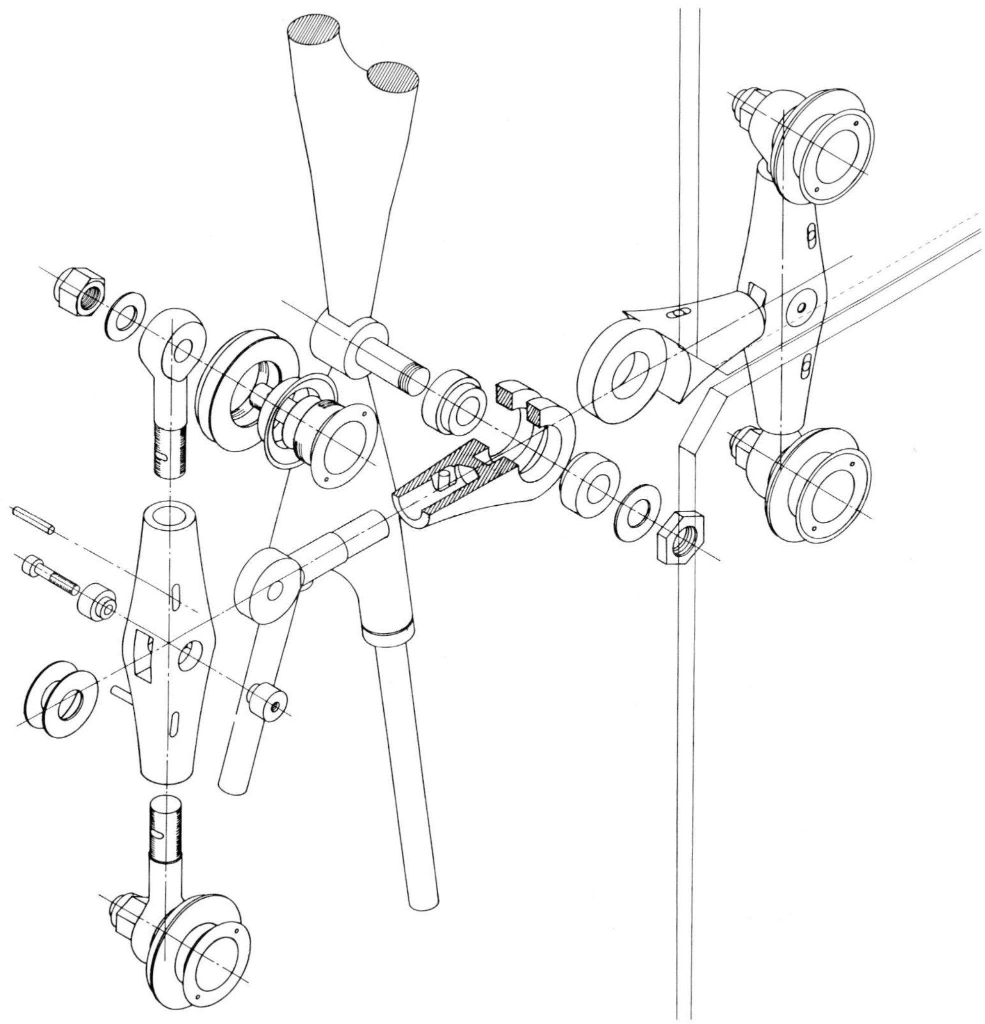

Exploded axonometric of the detail for fixing the glass, with the beginning of the vertical cable truss system and the chamfered corners of the glass sheets.

THE GREENHOUSES OF THE PARC CITROËN, PARIS

Date: 1992. Architect: Patrick Berger. Client: Ville de Paris, Département de Parcs et de Jardins. Building control engineers: Véritas. Contractor: Eiffel, large greenhouses; Sitraba, serial greenhouses. Cost per square metre of glazing (including structure) FF 6900 excluding tax, large greenhouses; FF 5600 excluding tax, serial greenhouses (1990 values).

The greenhouses (*serres*) of the Parc Citroën Cevennes, in the 15th *arrondissement* in Paris (completed in 1992), exploit to the full the transparency potential of the structural glass system developed at La Villette because the buildings, completely free-standing, can be seen through from one side to the other.

The park is a recent creation by the Paris Town Hall Parks and Gardens Department, on a site previously occupied by the Citroën car factory. Two successful teams were nominated following a competition held in 1985: Alain Provost (landscape), working with Jean-Paul Viguier and Jean-François Jodry (architects), and Gilles Clément (landscape), in association with the architect Patrick Berger. The latter's design for two large greenhouses ('*serres monumentales*') and six similar 'serial' greenhouses ('*serres sérielles*') were a key part of the overall park design.

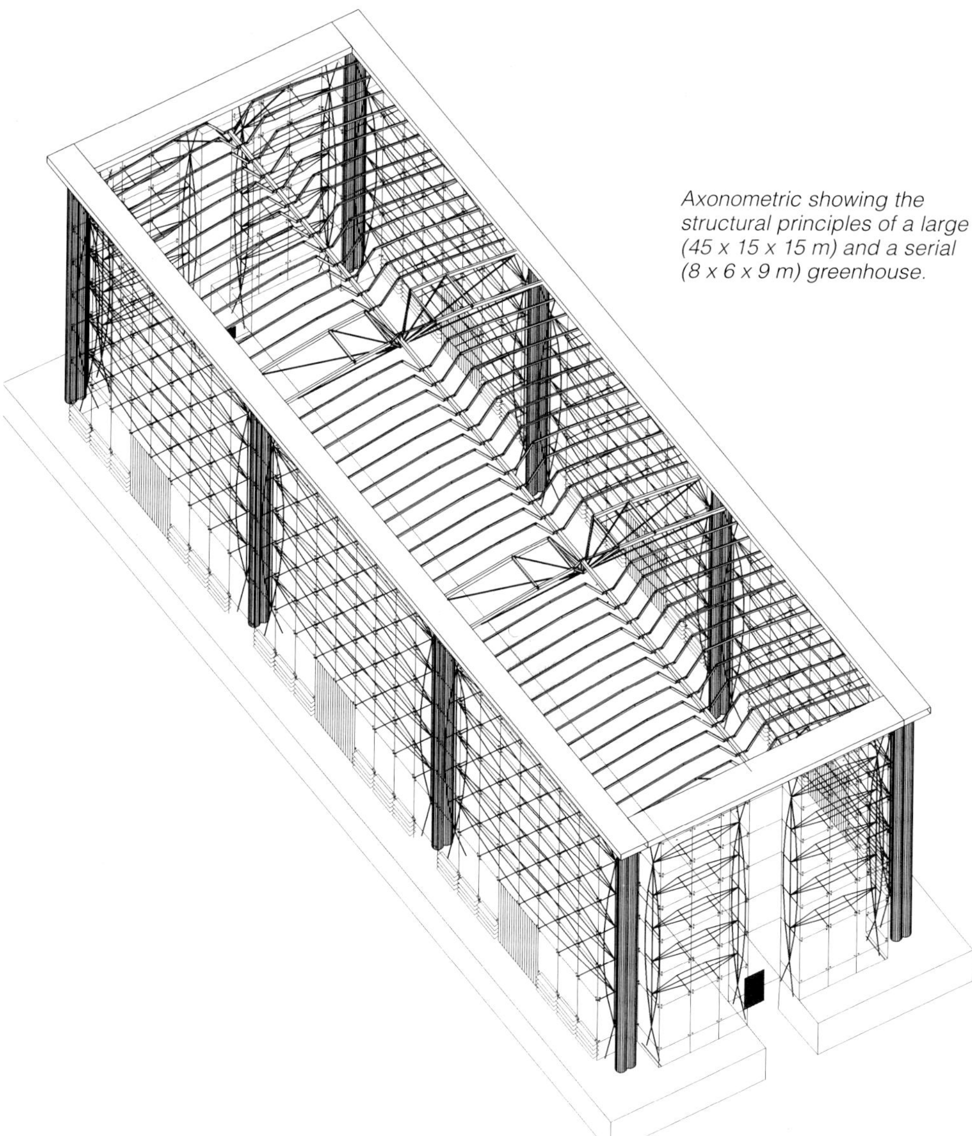

Axonometric showing the structural principles of a large (45 x 15 x 15 m) and a serial (8 x 6 x 9 m) greenhouse.

The architects' design philosophy for the greenhouse is based on an opposition between antiquity and modernity. Largely through his choice of materials he evokes a dialogue between the old and the new. Stone, bronze and wood are juxtaposed with stainless steel castings, cable tension elements, glass and silicone.

The two large greenhouses, called the *Orangerie* and the *Serre tempeñé*, are identical, as are the six serial greenhouses. They are rectangular volumes measuring respectively 45 m long × 15 m wide × 15 m high and 8 m long × 6 m wide × 9 m high. They are based on a simple structural frame with an orthogonal

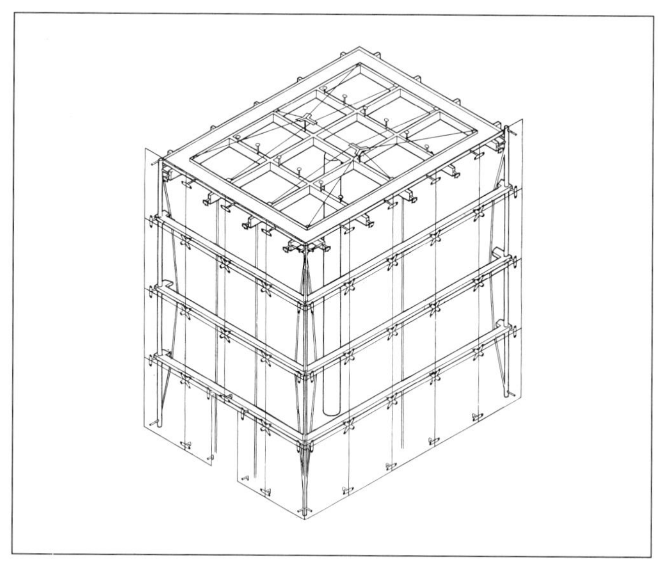

geometry. In the large greenhouses, the portal frames have 15 m high columns spaced 15 m apart and connected at their heads by a peripheral beam. This column and beam frame then supports the roof. For the serial greenhouses there is one central column with a steel pergola crowning it. The pergola is stabilized by vertical cables in the middle of each face.

These massive structural frames are clad in wood and stone, thus symbolizing the archaic components of Berger's compositions. The glass envelopes are fixed to the frames as discretely as possible.

The glass in the large greenhouses is suspended from the peripheral beam on wing springs, as at La Villette. It is also braced by cable trusses in a similar way, but with the exception that the cables form a two-level hierarchy of primary vertical trusses and secondary horizontal ones fixed to the verticals for each 15 × 15 m bay. The vertical trusses have a central spine compression member that resists the tension in the cables. These spine members are fixed to the spacer struts of the vertical cables. The tendency of the horizontal trusses to buckle vertically out of plane is resisted by small cables tied to the perimeter beam at the top and the floor slab at the bottom.

For the smaller serial greenhouses the glass is also suspended from the roof level, each face of the greenhouse being an independent panel. The wind loadings

Front views of a large and a serial greenhouse. The structural columns are clad in wood.

The large greenhouses have a double system of cable trusses; the vertical primary elements are fixed to the main structure; the secondary horizontal elements have a central compression tube. The serial greenhouses are made rigid by vertical cable trusses that also have a compression member at each of their corners.

on the glass are resisted by simple rectangular section beams fixed to the vertical cable trusses in each corner. Like the horizontal trusses of the large greenhouses, these vertical trusses have a compression member that resists the prestress of the cables.

In both cases, the glass detailing is identical to that developed at La Villette, with the exception of the four-point assemblies which, instead of being H-shaped, are X-shaped 'star' assemblies. This modification was introduced primarily for architectural reasons, but also because of the realization that each panel of glass volumes, once glued together with the silicone, is relatively rigid and therefore the 'racking' moments are quite limited. Instead of being resolved by hinging and geometric deformation, movement is allowed in the sliding of the bolt connections to the star-shaped casting on PTFE washers.

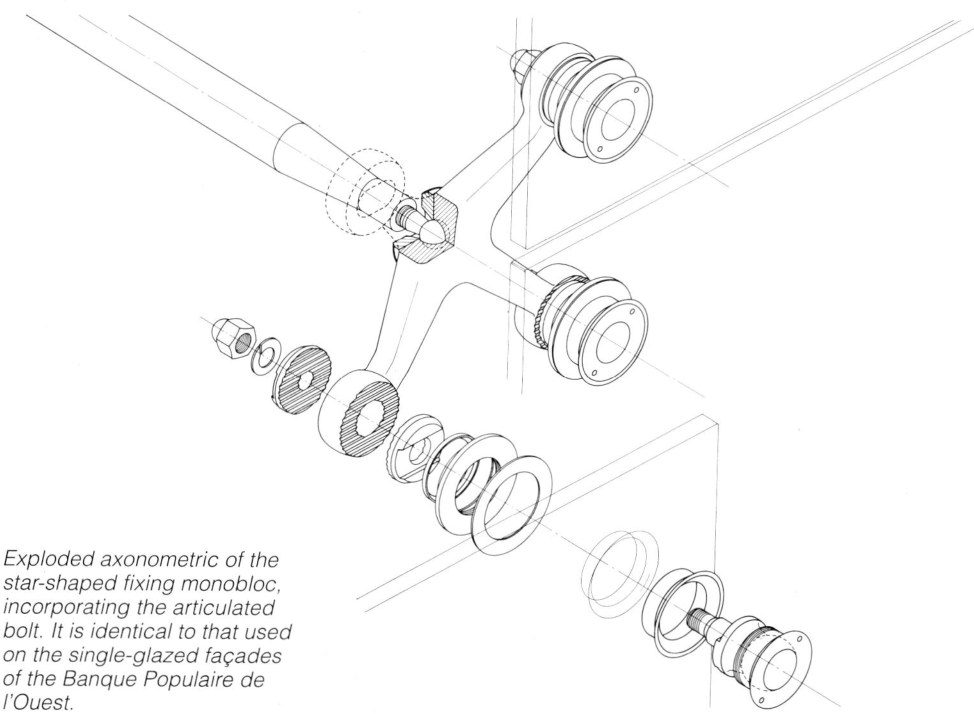

Exploded axonometric of the star-shaped fixing monobloc, incorporating the articulated bolt. It is identical to that used on the single-glazed façades of the Banque Populaire de l'Ouest.

DEVELOPMENTS IN JAPAN

Since the late 1980s Asahi Glass in Japan have been analysing and exploiting the articulated bolt glass system developed at La Villette. Asahi's principal concern is the behaviour of the system in seismic and typhoon conditions. Sophisticated testing is being conducted to analyse its performance in these extreme conditions (winds up to 350 kg/m^2 and seismic loading up to 1 G). Interestingly, the system may be used in such conditions without modification, apart from changes in the dimensions of the volumes of glass and in cable truss design to accommodate the more severe wind loading than that experienced in Europe.

An interesting conclusion of Asahi's laboratory experience is that in exploratory testing, suspended glass panels similar to those employed at La Villette could be sealed completely around the perimeter with mastic silicone (instead of using the extruded silicone movement joint) without excessive stresses or breakage occurring. Shear of up to 50% of the width of a mastic joint occurred, showing that a large part of the energy can be dissipated in the silicone.

Three projects by Asahi are illustrated here.

The first is an interesting application of sand-blasted toughened glass. The luminous effect of the small

At the foot of the tower of the Long Term Credit Bank in Tokyo (completed in 1993 by Nikken Sekkei Ltd), the entrance hall is surrounded by a façade of structural glass. Its impressive proportions required the use of a reinforced metal structure.

The translucent glass façade of a club in Tokyo, built in 1992 by Edward Suzuki, screens from view a small private courtyard.

translucent glass surface (180 m²) creates a subtle intimacy in the courtyard of this 1000 m² club, built in 1992 in Tokyo by the architect Edward Suzuki.

The second project is a 60 000 m² office block, built by Nikken Sekkei Ltd for the Japan Long Term Credit Bank not far from the Imperial Palace in Tokyo. The huge entrance hall is surrounded by a façade in structural glass inspired by those of La Villette but on a much larger scale (1800 m²): five spans in height, four spans in width and two in depth, each comprising nine (3 × 3) glass units (whereas La Villette counts 4 × 4 units). With such a layout the behaviour of the cable trusses is made simpler. A non-linear calculation is thus avoided, but the visual presence of the structure is stronger.

The third application of structural glass is a display window façade, 266 m², in front of a ship exhibited in the maritime museum, designed in 1994 by the architect Tetsuo Furuichi in Nagasaki. The cable trusses are of the same type as those of the Parc Citroën: they are fitted with a compressed central tube, but are laid vertically and not horizontally. According to the architects, the use of this device, which recalls a ship's wind-braced masts, is very appropriate within a building complex dedicated to the nautical environment.

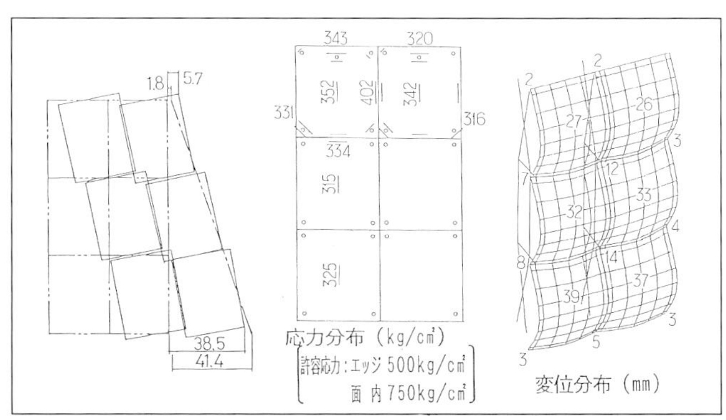

The tests to establish the behaviour of structural glass under seismic and typhoon loads were carried out by Asahi Glass.

The maritime museum in Nagasaki (designed by Tetsuo Furuichi in 1994) shows, from both the inside and outside, one of its main exhibits by a display window façade of structural glass. The central member, compressed by vertical cable truss systems, recalls the wind-braced mast of a boat.

50 AVENUE MONTAIGNE, PARIS

Date: 1993. Architects: Epstein, Glialman and Vidal. Client: Arc Union. Building control engineers: Véritas. Contractor: Eiffel. Cost per square metre of glazing (including structure, glass and maintenance) FF 11 300 (1992 values).

The Avenue Montaigne is one of the most exclusive addresses in Paris for a mixture of activities: fashion, offices, luxury goods. Arc Union, a large development company, undertook the transformation of several buildings behind the façade of 50 Avenue Montaigne into a multi-use commercial and office complex. At the heart of this new project the architects Epstein, Glaiman and Vidal decided on a large atrium and entrance hall as a way of bringing excitement and natural light into the complex. The client invited Peter Rice and the RFR team to design the glass atrium envelope to fit into the architectural context.

The atrium is south facing, giving on to a neighbour's courtyard. It is semi-circular on plan, to a radius of 8 m and 24 m high. The design is based on a response to this geometry. From the beginning, the idea of reflecting the half-cylinder in the glass, making an illusionary whole cylinder, was discussed with the architects.

This led to the creation of a structure with a radial geometry that reinforces the cylinder and it also meant that the inside surface should be as smooth as possible. The structure consists of a series of layers of cable trusses that are tensioned against the floor slabs every 3 m. The semi-circular geometry permits a very efficient configuration of one main cable, serving as a catenary for positive wind loads on the façade, pulled back to the slab with smaller radial cables to ensure stability for asymmetrical loading.

The glass pane, measuring the full 24 × 16 m, is suspended from an external bowstring truss. The countersunk faces of the articulated bolts are on the inside, thus creating a smooth reflective surface to mirror the semi-circular volume. The glass itself is fritted to create a source of diffuse reflected light at night.

Each volume of glass is suspended from the one above, as at La Villette, but with the star configuration of attachment piece developed for the Parc Citroën Cevennes and Banque Populaire projects. The stars are attached back to the cable planes by small stainless steel tapered struts, articulated at each end.

It was not possible to fix a typical cable truss at the lower end of the façade curtain, because the atrium

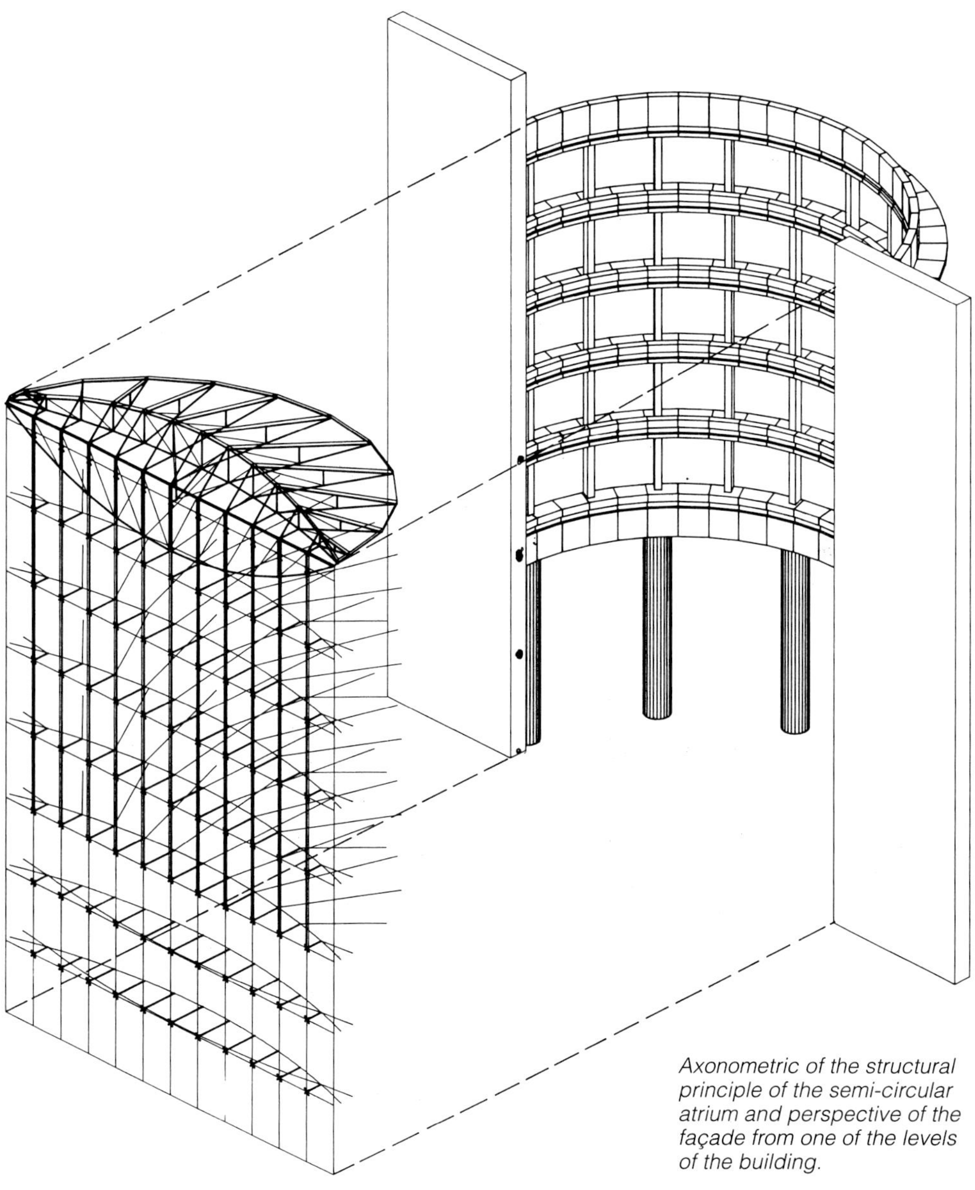

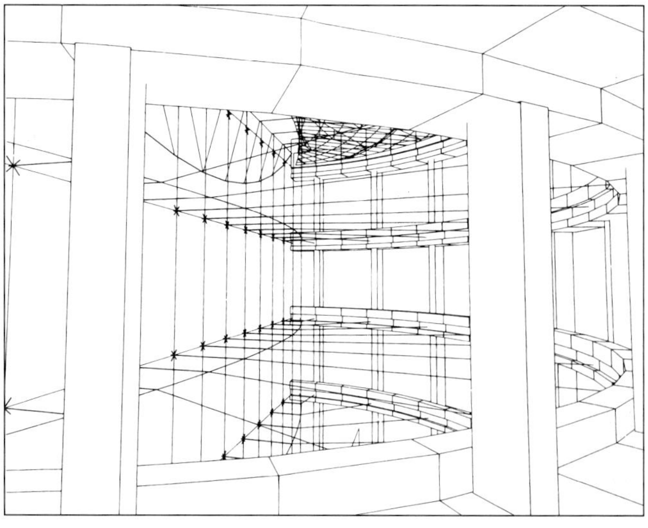

Axonometric of the structural principle of the semi-circular atrium and perspective of the façade from one of the levels of the building.

View from the courtyard. The 24 m high glass façade is suspended from a bowstring truss fixed on the exterior of the atrium. The fritting of the glass makes the façade a source of diffuse light at night.

The glass in the atrium is simultaneously transparent and reflective, giving a dual visual quality to the space. The façade allows the buildings that surround the interior courtyard to be seen, while reflecting the layers of cable trusses and the semi-circles of the atrium levels to create the image of a complete cylinder. At the base of the façade there is an independent cable truss system, with no radial cables.

ceiling is 7 m above ground level. Therefore an independent cable truss with a counter cable instead of radial cables was used, as on the La Villette trusses. The glass span is much greater than at the top levels of the façade. This was made possible using finite element calculations to determine the exact stresses in the glass. In the analysis it was apparent that relatively large deflections (up to 1/80th of the span) can easily be tolerated. Traditional approaches to glass calculations are based on limiting deflection in order to limit the concentration of stresses that occur at the fixing details. The articulated bolt allows considerable deflections without creating stress concentrations and therefore can allow relatively large spans. In this case, 3.8 m for 15 mm thick glass is the maximum. The final limitation on span is then governed by psychology: when it becomes uncomfortable for people to see large deflections of the glass plates, even though the volumes are stressed to perfectly acceptable levels.

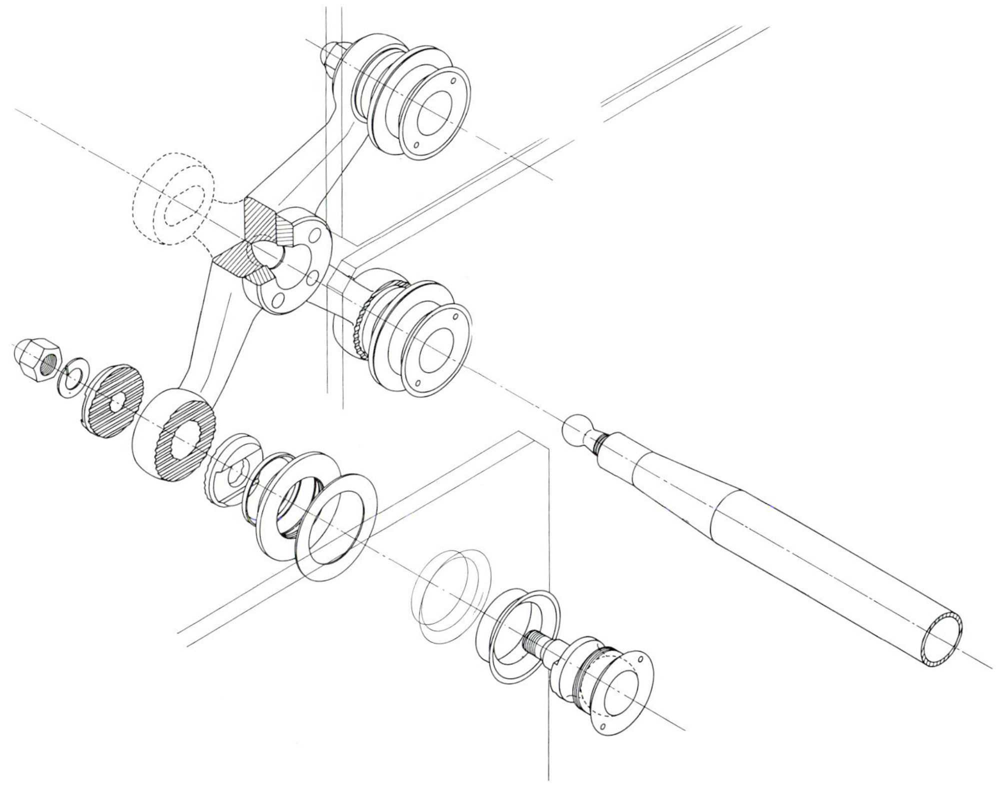

Exploded axonometric of the star-shaped attachment piece, similar to that used on the Banque Populaire de l'Ouest. The corners of the glass sheets are chamfered to accomodate the glass assembly pieces, which sit on the exterior, and the cable trusses, which cross the volume of the atrium on articulated struts of variable section that span the glass sheet.

The bare interior of the atrium's façade is completely smooth, as all of the fixing details are on the exterior, on the courtyard side. The glass pane is brought to life by the interplay of lines drawn by the cables and struts of the truss systems.

THE GRAND LOUVRE INVERTED PYRAMID, PARIS

Date: 1993. Architect: I.M. Pei. Client: EPGL. Building control engineers: Véritas. Contractor: Eiffel, structure; SIV, glass. Cost per square metre of glazing: FF 24 700 excluding tax (1991 values).

Peter Rice and RFR were invited by I.M. Pei to develop a second, inverted pyramid after having assisted his team on the Grande Pyramide in the final construction design and fabrication stages. This smaller pyramid is intended as a *lustre* (chandelier), reflecting daylight into the underground exhibition and shopping areas of the Carousel gallery. The inverted pyramid is more than a light source: it also a sculptural element in its own right.

The structure is almost entirely of cables, relying on the mass of the glass to maintain the tension. Each face of the pyramid is suspended from the concrete perimeter by its top edge. It hangs down as a curtain of suspended glass, in the same way as La Villette's suspended panels. These faces are then held in their inclined position by a series of small cables fixed to the middle of each lozenge-shaped volume. The cables are connected to a primary system of suspended 'flying struts' and main support cables.

Each volume of glass is attached to the one immediately above it by a cross-shaped star connection. The vertical bar restrains the dead weight and the horizontal bar restricts the inevitable pivoting of a volume around its vertical axis. Each connection to the glass is an adaptation of the spherical bearing articulated bolt developed for La Villette.

The roof glass, at the 'base' of the inverted pyramid, is supported by a completely independent network of cables. A distinct hierarchy of main cables spanning the entire width of the base and smaller secondary cables supporting the individual volumes is set up. The composition is based on a tartan grid, which facilitates the detailing by keeping the primary cable nodes separate from the secondary cable attachments.

The roof glass follows a very flat pyramid shape, to

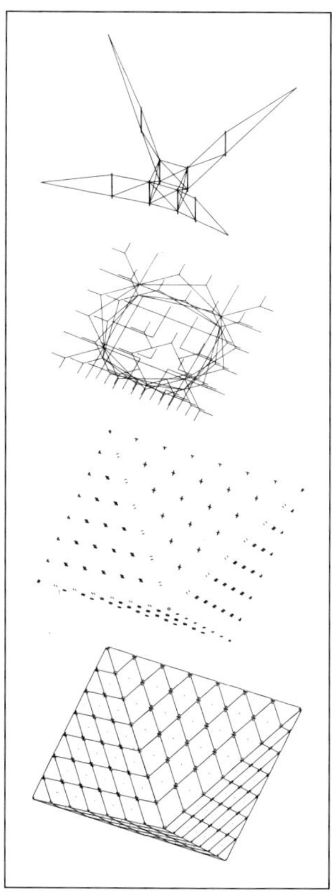

The elements of the pyramid. From top to bottom: primary structure; suspension system; glass panels.

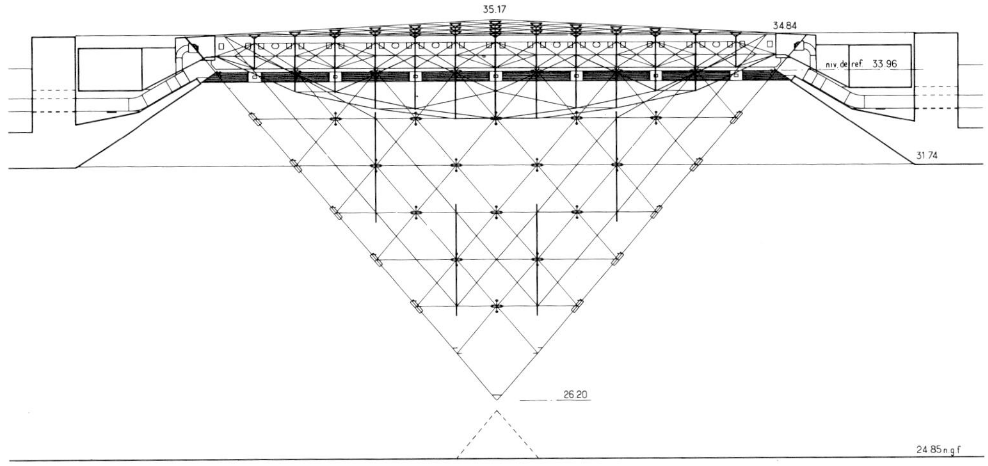

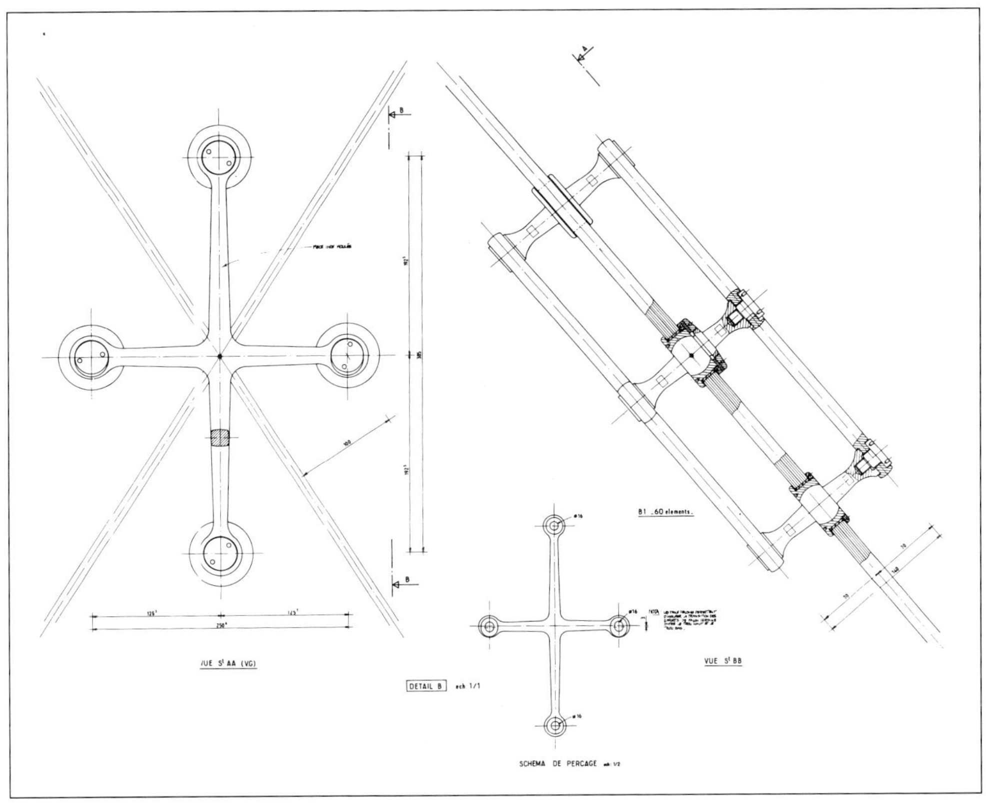

ensure rain water drains off. The volumes are in a triple laminate with chamfered edges capable of supporting an eventual, although unauthorized, load of people walking on the glass.

The inclination of the glass surfaces makes them particularly effective as light reflectors. There is less light coming from the surfaces behind the inclined glass than from the sky above, thus making them highly reflective. Large quantities of light are reflected in this way into the underground corridors, and the pyramid itself appears particularly luminous because of the relative light values.

Left and right:
General and close up view of glass suspension system.

Above:
Section of the pyramid. Each of the faces is suspended from the peripheral concrete structure. The roof, formed by the 'base' of the pyramid, is itself a flattened pyramid, to encourage drainage of rain water.

CHANNEL 4 HEADQUARTERS, LONDON

Date: 1994. Architects: Richard Rogers Partnership Client: Channel 4. Contractors: Eiffel; SIV, glass furnishings. Cost per square metre of glazing: ?

The focal point of the Channel 4 Headquarters building in London, designed by the Richard Rogers Partnership, is the spectacular glazed entrance hall. The wall is a curved suspended glass curtain forming a quarter-circle plan.

This curved surface is framed by two glass returning panels and topped by a sloping glass roof. This complex glass skin is suspended from an array of steel cantilever fingers at roof level and braced by a two-way cable net. A network of vertical and horizontal cables are tensioned against each other, forming an array of fixed points in space inside the glass plane. The glass is separated from these points by small articulated struts.

Predicting the performance of the cable system is particularly difficult, given the complex geometry, the high prestress values required because of the relatively flat curve of the net and the consequent problems of creep in the anchorages in the concrete building frame.

Right:
The entrance façade, situated in a corner of the building.

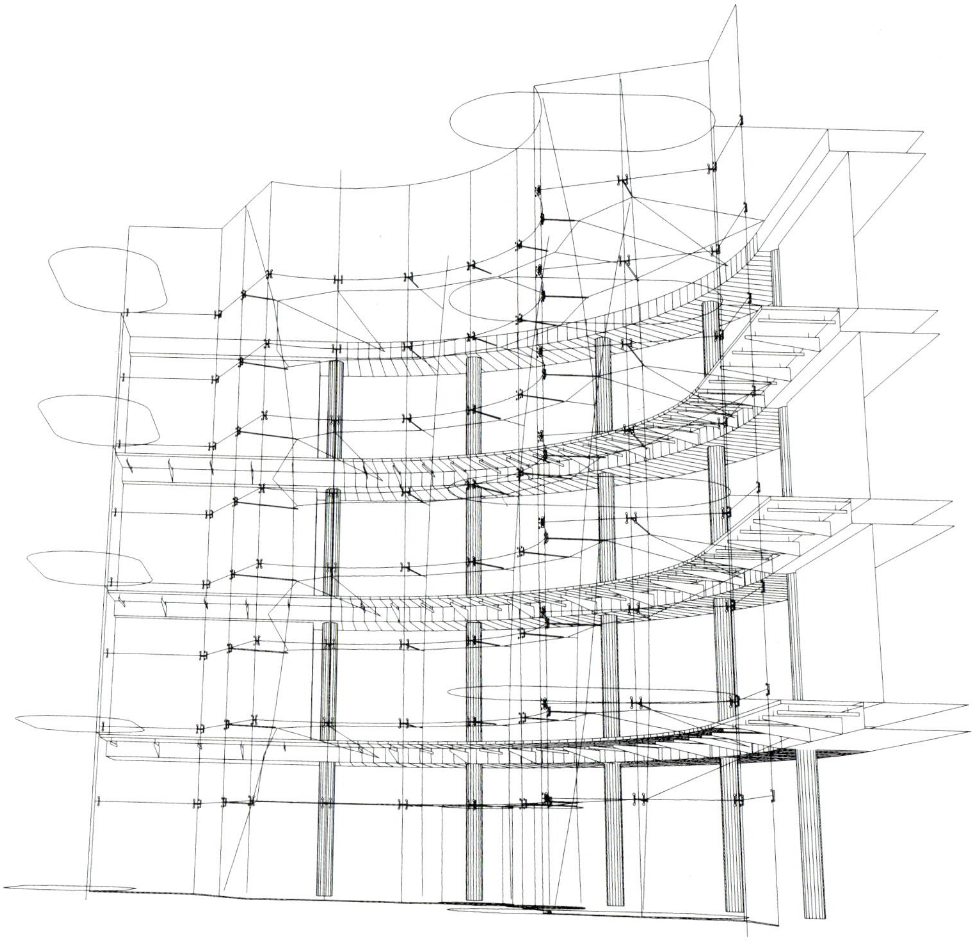

Below:
Perspective from the exterior of the façade, with the complex system of cable trusses.

The glass suspension system is conceptually the same as on the façades of the Serres at La Villette. It is hung in a chain from spring compensation assemblies, all of which are articulated to ensure that the load path can be clearly predicted and that no external forces can become attracted to the system.

The curved plan shape, however, requires adaptations of the system because the surface geometry is a form that can resist wind load – as an arch in suction and as a chain under positive pressure. It is therefore essential that all of the horizontal connections permit movement in the plane of the glass, so that the curve will not resist the load, but will instead transmit it to the cable structure designed to resist it.

Given the sensitivity of the system to relative movements, the silicone joints play an important role in the behaviour of the façade. Most of the joints are of a low modulus silicone, to allow the curved surface to

'breathe'. However, in the angled 'corners' the joints have a structural function. The glass panels on the edges forming the corner depend on the return panels for wind load stability. The return edge behaves as a continuous glass beam from the top of the façade to the bottom of this edge. The silicone is therefore required to resist the shear deformation of the truss in asymmetric loading conditions and is consequently specified as a high modulus product.

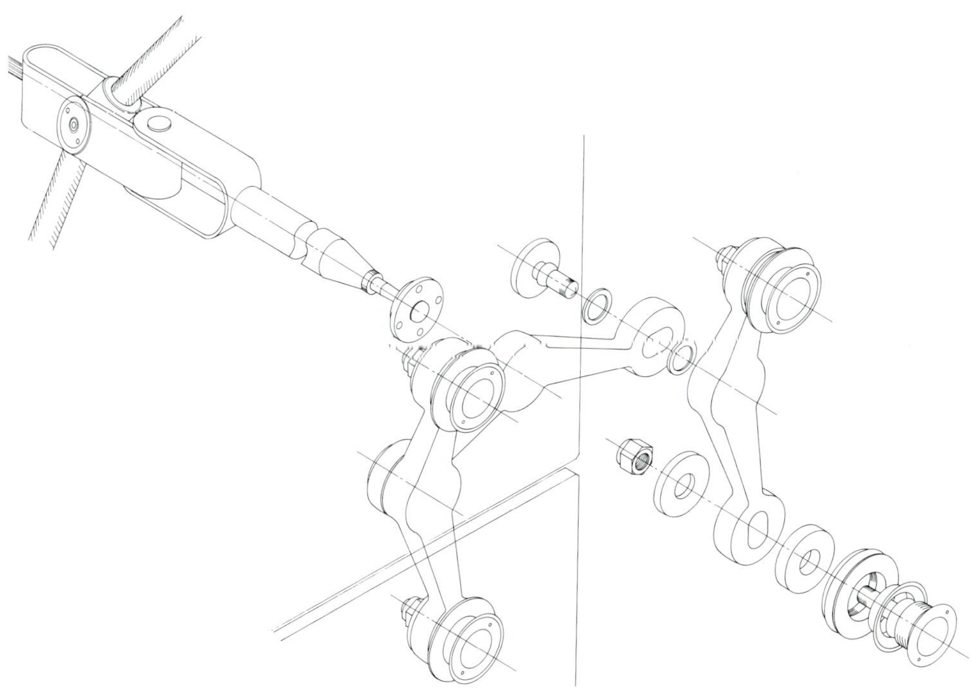

Above:
Details of the curved façade from the interior of the atrium. On the right, the vertical cable truss is aligned with the joint between the glazed panels.

Right:
Exploded axonometric of a four-point, H-shaped fixing component.

TGV RAILWAY TERMINAL, ROISSY

Date: 1994. Architect: ADP. Client: Aéroports de Paris (ADP) and SNCF. Building control engineers: CEP Véritas. Contractors: Watson, metal structure; Fischer, glazing. Cost per square metre of glazing: FF 3300 tax excluded (1991 values).

The Charles de Gaulle airport expansion programme includes a TGV railway station and passenger exchange complex serving Terminal 2 and the proposed Terminal 3, designed by the architects Paul Andreu (ADP) and J.M. Duthilleul (SNCF), with the engineering group RFR under Peter Rice.

The station straddles the main axis of the symmetrical Terminal 2 buildings underneath the exchange complex, which is crowned by a hotel building. It is covered by an all-glass shelter that includes lateral walls to give complete protection to the platforms. In all, the station is 500 m long, which is as long as the longest TGV train, and its glass roof and lateral wall covering use 28 000 m^2 of glass.

The architectural concept for the station is based on a cinematic of different views of this enormous glass covering that is revealed to travellers arriving and

departing by train. From the exchange complex, they can see into the roof tension structure and appreciate its metamorphosis from an elevated sloped shed over the escalators down to the curved covering of the railway trench.

The roof structure is designed to permit a clear view

The 500 m long station is positioned perpendicular to the main axis of the buildings of Terminal 2.
The side view expresses the options for different treatments of the glazed surfaces of the façades and of the roof. The latter is a translucent sheet, visually separated from its support, while the former are also as invisible as possible. Their respective structural grids are staggered: the croissant beams do not coincide with the exterior masts.

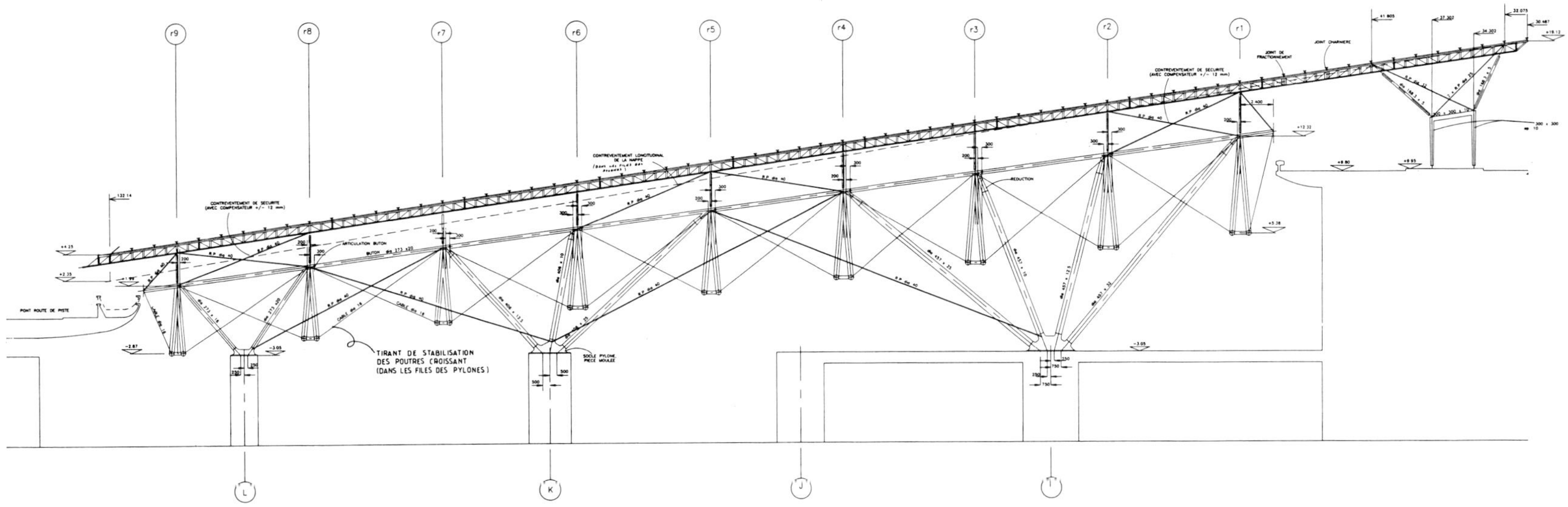

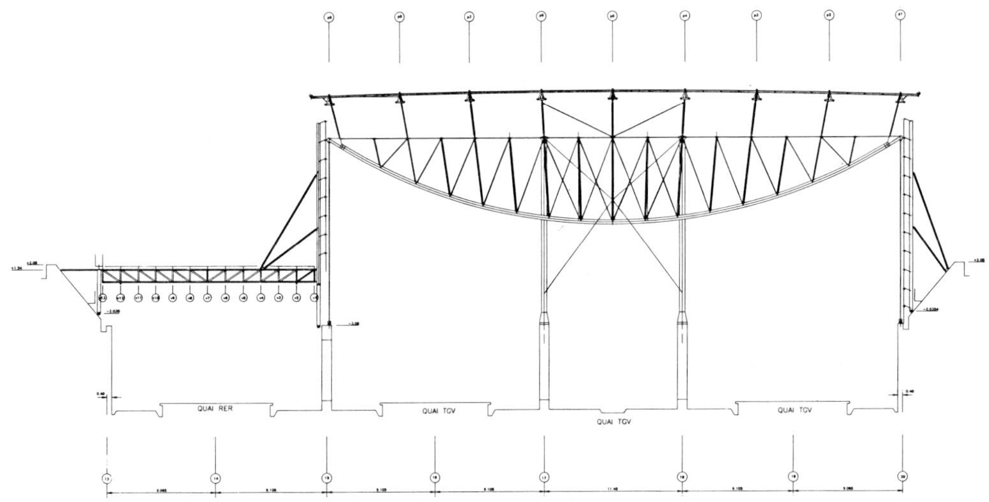

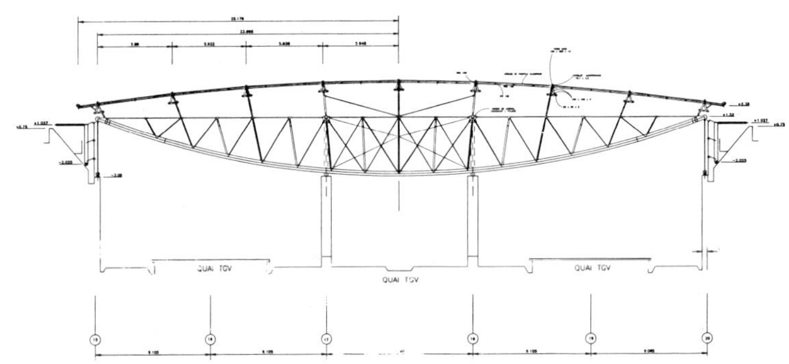

from the platforms of the surrounding airport. This is achieved using beams supported in the centre, their ends, with anti-uplift tie downs, acting as cantilevers. This configuration minimizes obstruction of the view out.

These cantilever beams, called 'croissants' because of their curved profile, support, on the end of small struts, a two-way grid of small steel beams to which the glass covering is fixed. Thus, the glass covering appears to 'float' above the croissants. The glass is fixed using a simple silicone gasket extrusion mesh and local hold down patch plates on the edges screwed on to the steel grid. The roof glass is fritted to reflect the sun's rays in the daytime and to serve as an indirect source of diffuse, reflected artificial light at night. This glass surface, with its steel supporting grid, is intended to be luminous and not disappear in transparency.

The lateral walls, however, are designed to be as transparent as possible, to maximize the view out and to encourage the feeling that the luminous roof surface is floating above. The clear, toughened glass is fixed to cantilever masts which are spaced 4.75 m apart on a grid staggered with the croissant tie downs; this configuration makes it clear that they are not holding up the roof. The masts are braced torsionally by a system of prestressed cables close to the plane of the glass. Compression members that would obstruct the transparency are kept to a minimum.

The glass panels are fixed individually to small cantilever 'arms' using the La Villette bolts and a sophisticated 'hand' complex in stainless steel castings. The hand system consists of individual 'finger' suspension brackets on the underside to support the glass panes. On the top side of the hand assembly is another series

Longitudinal and transverse sections of the station platforms.

AEROPORT CHARLES DE GAULLE TGV

The conception of the croissant beams, with vertical anti-uplift tie downs at their ends acting as cantilevers, allows optimum views of the aeroplanes from the interior of the station.

of smaller articulated fingers that restrain the glass against loads perpendicular to the surface. The fingers are articulated to reduce the rigidity of the linking between the panels and thereby allow differential movements such as thermal expansion, creep and deflections without loading the plane of the glass. This is particularly important given the size of the lateral wall, almost 100 m long, and the deflections at the top of the cantilever masts, which are as much as 17 m high. The hand assembly also includes a complex system of adjustment to compensate for errors in positioning of the arms and masts.

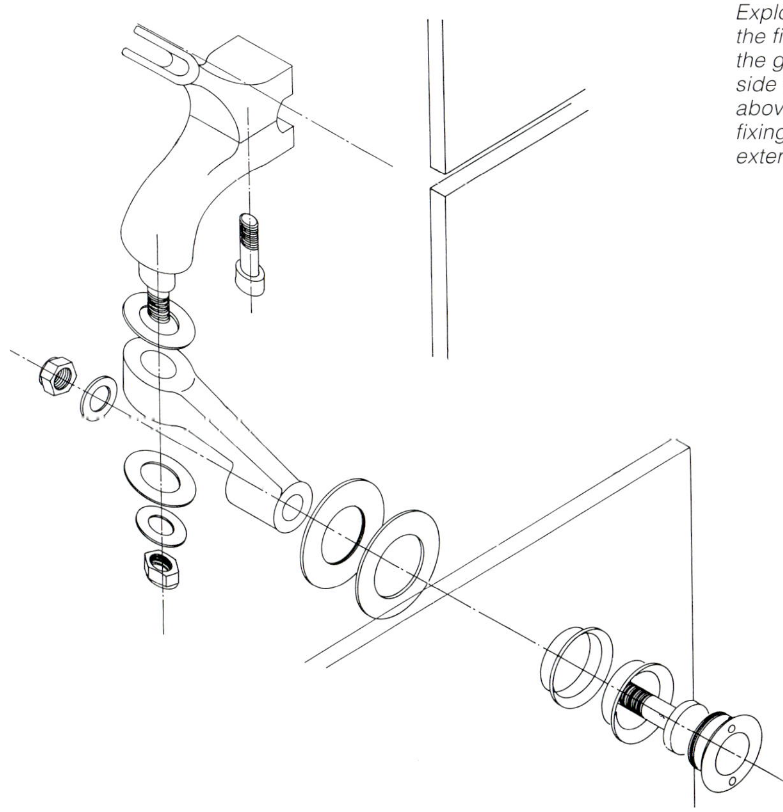

Exploded axonometric of the fixing component of the glass panels on the side façades with, above, the casting for fixing to the arms of the exterior masts.

THE WEST FAÇADE OF LILLE CATHEDRAL, LILLE

Design stage of the project in 1994. Architects: Pierre-Louis Carlier, RFR. Client: The Bishopric of Lille. Estimated cost per square metre of façade: ?

The Cathedral of Notre Dame de la Treille in Lille was constructed during the nineteenth century, but the west front, with its two towers, was left incomplete. The façade is currently clad in a temporary covering of brick and wood. The architect Pierre Louis Carlier, with the bishop Monseigneur Vilnet and the Association Diocésaine de Lille are putting together a project to finish this western front.

The design team includes the artist Ladislas Kijno with his team of stained glass window craftsmen Gruber, Pion and Loup for the rose window and Peter Rice with RFR for the rest of the façade.

This project is still in the preliminary design phase, but represents a particular development in the structural glass design philosophy and method in its exploitation of the potential expression of a luminous plane of glass.

Kijno christened the façade project the 'Veil of Véronique', a reference to the Christian legend of Véronique, who used her translucent veil to wipe away the blood and sweat on Christ's face as he carried the cross to the place of crucifixion. According to the legend, stains describing Christ's face remained and are still visible on the veil, which is kept as a Holy relic in Italy.

The project involves a series of ogival arches that correspond to the sections of the nave and the two side aisles. Each arch is realized in prestressed 'Pierre Bleue', a very hard, local limestone, using a network of cables and struts inside it. This stone was chosen because the existing cathedral is largely clad in it.

Stone worked by modern technologies is consciously used for its symbolic nature. The design of ancient cathedrals relied explicitly on the stone masonry techniques of the time at which they were built: the Lille design relies on this philosophy in the same manner.

The 'veil' itself relies largely on stone and modern technology. It is conceived in suspended translucent marble. Marble was chosen because of the specific light quality required inside the cathedral. The colour must correspond with Kijno's work for the rose window and must also allow for the relatively low light levels inside.

Perspective of the façade, with suspended translucent panels and Ladislas Kijno's rose window.

The projects we have just illustrated bear witness to the development of ideas and the design process established for the Serres at the Cité des Sciences et de l'Industrie at La Villette. These projects may belong to very different contexts but they all work towards a greater appreciation of glass. Indeed, we have yet to exploit fully the structural capacity of this material. Glass is very fragile but becomes quite resistant when its natural capacities are exploited by the way loads are applied. Furthermore, it possesses remarkable architectural qualities. By virtue of its transparency and the natural duality of its appearance – between the material and the immaterial – it allows the definition of surfaces and of subtle and ambiguous spaces.

Ten years have passed since the project at La Villette. More recent projects using the articulated bolt suspension system confirm its potential for increasing transparency. The commercial exploitation of the system has been assured following the filing of a patent: in Europe, by SIV (Societa Italiano Vetro) belonging now to the British company Pilkington, the Fischer Company being its main distributor in Germany. In Asia, the patent is exploited by the Asahi company. A close cooperation with these companies is allowing the techniques and their application to an industrial process to be improved. It is also assuring the control of patent infringements by companies wishing to produce economical versions of the system without fully understanding its technical implications. Indeed, the use of structural glass necessitates the incorporation of a range of very specific devices to protect it at the points at which stresses are the greatest. Such devices have emerged following lengthy research and in depth testing and therefore should not be modified without further research.

The future of glass in building is closely related to the climatic role it plays when used in façades: it can either help energy loss or energy gain, depending on its exposure to the sun. Its low thermal insulation coefficient means that it can easily capture solar energy. Intensification of research work on the use of glass for insulating double glazing is now essential, both to limit energy losses and to control solar input.

Finally, the message we wish to convey through this book concerns the importance of reflection on how a material – here glass – is made to work and to the themes of process, predictability and hierarchy. It is only with an awareness and understanding of the design method, applicable to any project, that one will be able to create the confidence necessary to guarantee the success of a project.

Photographic credits

The technical drawings reproduced in this book were all specially drawn by RFR.

ADP: p. 141.
Henry Bardsley: pp. 22, 135.
Jocelyne van den Bossche: pp. 50, 56.
Luc Boegly/Archipress: pp. 116, 117, 118 r., 134.
François-Xavier Bouchart/Archipress: p. 126 r.
Jean-Louis Cohen: p. 13 t.
Peter Cook/Archipress: p. 137.
Stéphane Couturier/Archipress: pp. 6–7, 9, 18, 19, 39, 58, 64, 91, 115 r., 118 l.
Christophe Demonfaucon/Moniteur: pp. 120, 121.
Michel Denancé: pp. 122, 132, 133, back cover.
Michel Denancé/Archipress: pp. 23, 138 b., 140.
Guy Deshayes: p. 41 br., 60 br., 66.
John Donat: pp. 44 t. and bl.
Dumage/Studio Littré: p. 8.
Hugh Dutton: pp. 38, 40 l., 59, 73, 74 b., 86, 88 l., 92 m and bl., 95, 96, 115 l., 119.
Kathuaki Furudate: pp. 20, 127, 128, 129.
Giraudon: pp. 10, 11 t.
Alain Goustard/Archipress: cover, pp. 17, 24, 27, 40 r., 41 t. and bl., 60 except br., 61, 82, 87, 88 r., 90, 92 t. and br., 93, 111 r.
Pino Guidolotti: p. 111 l.
G. Halary/ADP: p. 138 t.
Ken Kirkwood: p. 44 br.
Ana Maksimiuk: p. 131.
Jean-Marie Monthiers: pp. 28, 98.
Eamonn O'Mahory: p. 136.
Pinster & Tahl: p. 113.
Etienne Tricot: p. 74 t.
Fernando Javier Urquijo: pp. 21, 124, 125, 126 l.
D.R.: pp. 11 b., 12, 13 b., 14, 15 l., 16, 65.

INDEX